Repair & Renovation
of Modern Buildings

Repair & Renovation of Modern Buildings

Ian Chandler

McGraw-Hill, Inc.
New York St Louis San Francisco Auckland Bogotá
Caracas Mexico Lisbon London Madrid
Milan Montreal New Delhi Paris
San Juan São Paulo Singapore
Sydney Tokyo Toronto

© Ian E. Chandler 1991

First published 1991

All rights reserved. No part of this publication
may be reproduced, in any form or by any means,
without permission from the Publisher

Typeset by Lasertext Ltd, Stretford, Manchester
Printed in Great Britain by
Courier International Ltd
East Kilbride, Scotland
for the publishers
B T Batsford Ltd
4 Fitzhardinge Street
LONDON W1H 0AH

Library of Congress Cataloging-in-Publication Data
Chandler, Ian E.
   Repair & renovation of modern buildings / Ian Chandler.
      p.   cm.
   Includes index.
   ISBN 0-07-011030-1
   1. Buildings–Repair and reconstruction. 2. Buildings–Remodeling
for other use. 3. Concrete construction. I. Title. II. Title:
Repair and renovation of modern building.
TH3401.C42   1992
690'.24–dc20   92-9893
                                                                                       CIP

Copyright © 1992 by Ian Chandler. This edition published in North
America by McGraw-Hill, Inc. All rights reserved. Except as permitted
under the United States Copyright Act of 1976, no part of this
publication may be reproduced or distributed in any form or by any
means, or stored in a database or retrieval system, without the prior
written permission of the author.

ISBN 0-07-011030-1

CONTENTS

Acknowledgment   8

1 LET'S REFURBISH

Phase 1 Inter-war non-traditional stock
   1919–1944   10
Phase 2 Post-war low-rise stock
   1945–1955   11
Phase 3 1960 to early 1970s   11
Current technologies   13
The refurbishment process   14

2 PROBLEMS, PROBLEMS

Structure   20
Defects   21
Concrete creep and shrinkage   21
Carbonation   21
Chloride attack   22
Corrosion of the reinforcement   23
Insufficient steel reinforcement   23
Poor fixings between structural elements   24
Spread of loadings   24
Corrosion of ties and fixings   24
Misalignment of precast concrete panels   25
Inadequate movement joints between claddings and structure   26
Poor joint design   26
Inadequate site quality control   27
Degradation of components   27
Low standards of insulation   28
Surface finish failure   28
Distortion of panels   28
Problems with other building structures   29
   Steel framed buildings   29
   Timber framed structures   29
Social influences on building defects   30
Summary   33

3 DIAGNOSIS

Phase 1 investigations   36
Phase 2 investigations   36
Phase 3 production information for remedial works   36
Methods of investigation   36
High rise buildings   36
The Bison wallframe system   38
   Brief description   38
   General findings   40
   Chemical attack   40
   Mosaics   40
   Joints   40
   Panel fixings   40
   Ties   41
   Summary   41
Framed steel clad two-storey houses   42
   Characteristics   42
   Problems   42
Reema hollow panel house   42
   General description   42
   Investigation   42
   General points   44
Timber framed houses   44
   External inspection   44
   Internal inspection   45
   Cavity inspection   46
Steel framed structure with lightweight cladding   46
   Access and equipment   47
   Visual inspection   47
   Testing   48

Other tests   *48*
Further investigations   *50*
Services investigation   *50*
Gas   *50*
Fixtures and fittings   *50*
Summary   *51*
Recording inspections   *51*
The log book   *51*
Computers   *52*
Key abbreviations and gradings used on computer print-out   *53*
Summary   *55*
User evaluation   *55*
Building in place   *55*
Appraisal   *55*
Desk study   *55*
Local authorities   *56*
Tenants   *56*
Limitations of records   *56*
Site inspections and testing   *56*
Stages for inspection and testing   *57*
Cracking   *57*
Quality of original construction   *57*
Safety   *57*
Collapse   *57*
Laboratory analysis   *58*
Types of test   *58*
Expected behaviour of the structure in service   *58*
The Report   *58*
Writing the Report   *58*
The body of the Report   *58*
The conclusions   *59*
Summary   *59*

4 OPTIONS

Emergency   *61*
Long term   *62*
Medium term   *62*
Short term   *63*
Assessment of physical condition   *63*
Criteria for options   *64*
Life cycle costing   *66*
Value engineering   *69*
Life of materials   *69*
Failure of remedial solution   *72*
Performance specification production   *73*

Performance requirement   *73*
Performance criteria   *73*
Evaluation techniques   *73*
Performance specification   *73*
Intelligent buildings   *74*
Factors in optimal solutions   *74*
The effect of housing management on options   *75*
Summary   *77*

5 DOCUMENTATION

Initial inspection report   *78*
Contract   *79*
Safety   *79*
Material sampling and analysis   *79*
Full structural report   *80*
Option analysis   *81*
Brief for remedial works   *82*
Systems needing minimal pre-contract information   *83*
Sources of information   *83*
Content of document   *83*
Suggested draft clauses   *85*
Basis of calculation of final payment   *85*
Clauses: standing time of operatives   *86*
Payment for repairs   *86*
Suggested schedule of rates for access cradles   *87*
Competition   *87*
Forms of contract   *87*
Elemental repairs   *88*
Type of documentation   *88*
Sources of information   *88*
Content of document   *88*
Competition   *89*
Forms of contract   *89*
Safety   *90*
Short term repairs   *90*
Type of documentation   *90*
Sources of information   *90*
Content of document   *90*
Elemental bills   *90*
Repairs   *90*
Method of preparation and finishing of concrete   *91*
Form of contract   *92*

Long term repairs   92
   Type of documentation   92
   Sources of information   92
   Content of document   92
Concrete repair   92
   Documentation for high quality concrete repairs   95
   Low-rise PRC houses   97
Replacement of unstable brickwork   98
Replacement of external brick panels   99
Applied external finishings   99
Documentation generally   99
Forms of contract   99
Safety   100
Summary   102
Programme plan and intentions   102
Record of work done   102
Valuation records   103
Log book   103
Summary   104

6 COST FACTORS

Comparison of methodologies   105
   Access   105
   Sequencing   108
Comparison of materials   108
   Costing and tendering   109
Pricing strategies by contractors   112
Site visit   114
Liaison with consultants   114
Costing examples   116
Summary   118

7 SITE ORGANISATION

Occupied buildings   119
Programming   123
Interpretation of documents   123

Quality Assurance   123
Safety   127
Examples of site methodologies and sequences   128
   Low rise   128
   High rise   131
Summary   133

8 TYPICAL SOLUTIONS

Study One: Bison cross wall flats in Walsall   134
Study Two: Two 16 storey blocks of flats in Stafford   138
Study Three: Four storey LPS maisonettes, Hull   142
Study Four: Wates house reinstatement   147
Study Five: Les Fenetres de Balzac – a 16 storey block of 320 flats in La Courneuve   149
Study Six: Residence Chateaubriand – 3 blocks of 160 flats   151
Study Seven: Houses at Gravelines, Pas du Nord   153
Study Eight: Apartment building, Chicago, USA   154
Study Nine: Office refurbishment, Edgware Road, London   157

9 IS IT WORTH IT?

Cost evaluation   161
Energy evaluation   164
Social concerns   164
Warnings   166
Summary   168

References   170

Index   172

Acknowledgment

This book has grown from the work of many contributors, both academic and professional who were initially involved with the TERN Project (Training for the Evaluation and Repair of Non-Traditional Buildings). This was funded by the UK government to produce distance learning materials, videos and handbooks for those people engaged in the complex problems of assessing, maintaining and repairing defective buildings, mainly housing. Therefore initial acknowledgment goes to the authors of the handbooks and the team who produced the five videos and twenty six handbooks within a very short time. Special mention must be made to David Burns, the Project Director, who brought it to successful fruition; also the City of Birmingham housing department's technical officers and Bullock, Tarmac, Kendrick building contractors who contributed valuable time, effort and expertise. This book's foundation in practical procedures and solutions is due solely to their contribution. My work with the Building Research Establishment, and their excellent research and subsequent publications in this field, has been consolidated and extended by further visits to organisations and repair/refurbishment projects in Europe and the USA. My wife, Ellen, accompanied me on many occasions, and her presence allowed me to enter households which would have been naturally suspicious of a man with questions and a camera. I'm grateful for the insight she provided in giving a human face to structural problems.

A list of organisations and people who have provided information is insufficient to record thanks, but here it is; Wandsworth Borough Council; Chicago Housing Authority; Slovak Technical University; Lyster Grillet and Harding; John Hunt; National Federation of HLM's, France; Walsall Metropolitan Borough Council; Glasgow City Council; Bristol City Council; Laing Construction; Jack Dilworth; Sally who typed this and Thelma who edited and shaped the book.

Interpretations, comment, opinion are my responsibility and are not attributable to any organisation or person.

Birmingham 1991   *Ian Chandler*

CHAPTER ONE

LET'S REFURBISH

The intention of this book is to provide an extensive overview of the many issues and technologies that have to be addressed and applied in the refurbishment of modern buildings. It is necessary to define modern buildings in this context. Broadly, it is those buildings constructed in concrete, steel or timber using either a frame or panel structural system. The emphasis will be on concrete structures, both in situ and precast as these are the most numerous and also tend to provide the greatest problems. The problems are especially acute in housing, where a boom in systems erected in the 1960s and early 1970s has provided a legacy of defects and low standards. Unfortunately the problems are not confined to housing. Hospitals, offices and schools constructed in systems are exhibiting major defects. Although much publicity, and research, has been given to large panel systems (LPS) and precast reinforced concrete (PRC) structures there are also similar problems with in situ concrete frames with a range of claddings. Currently, at least in the UK, there do not appear to be many structural problems with steel framed structures, probably because there are not many so constructed, and the majority of these are recently built. In the USA it is the norm to construct in steel. Those high rise buildings erected early in the twentieth century are still performing adequately with no signs of structural problems. In common with European experience, some in situ concrete framed buildings are showing degredation of the exposed concrete frame.

The history of prefabrication and system building is very carefully explored by RUSSELL (1981). There are three main influences which have encouraged the use of system building, with its flirtation with off-site factory production and use of mechanised assembly processes. These are:

- architectural stylistic movements
- need for large numbers of buildings within a short space of time
- desire to increase productivity and profitability.

Architectural rationales, coupled with social aspirations in the 1920s and 30s saw the idea of a house as 'a machine for living in' (LE CORBUSIER) to be a reality. If form, space and structure could be determined to meet living needs, then by reducing these to a simple layout would lead to the means of mass production. This is reflected in factory systems, where function of production process would determine the output of mass products.

In the USA RUDOLF SCHINDLER, architect, experimented with prefabricated concrete panels for his own house in 1921/22. These storey height panels were 1 m wide. A further experiment with modular grids produced, in 1947, the 'Schindler Frame' for low rise timber framed houses. An early problem in one of SCHINDLER's designs for a housing development showed that single skin concrete walls could not cope with extremes of weather – even in Southern California.

Another American influence in the prefabrication and modular concept was ALBERT E. BEMIS. His third volume 'Rational Design' (1933), in the series 'The Evolving House' foresaw the mass production of house parts. This was based on a cubicule model interlinked into a matrix of three dimensions.

In the twentieth century the architectural ideas have coincided with two major world events, and social aspirations throughout the western world, producing a third. Two world wars created the need to replace and upgrade the building stock. This was not necessarily in response to war damage, but the desire to move forward, after conflict in the provision of better living and working environments. In the late 1950s, in the UK, there was a national political debate concerning the standards and numbers of housing. It was a major governmental electoral issue and voters saw it as a high priority. The two main political parties vied with each other in promises to build more and better public housing. This resulted in the third event which was system building.

PHASE 1   INTER-WAR NON-TRADITIONAL STOCK 1919–1944

Between 1919 and 1944, 52,000 non-traditional houses were built in the UK, using many traditional features and internal floor plans. They looked, internally and externally, similar to lead bearing solid brick wall houses, but the external walls were replaced by a combination of concrete blocks or in situ and pre-cast concrete frames and cladding components. Also used were steel frames with a combination of infill panels, for example, 10,000 Dorlonco types. Over 8,000 concrete block Boot houses were constructed nationwide. Floor space and estate layouts were generous and attractive. Services and facilities were the best available.

From this phase over 25% of house types were eventually designated by government as being defective and therefore came within the regulations of a scheme providing grant assistance for major structural reinstatement repairs.

1.1   PRC Cornish with reinstated house in load bearing brick/block walls. *Bristol*

PHASE 2   POST-WAR LOW-RISE STOCK 1945–1955

After the war a combination of factors made the concept of system building attractive. These were:

- lack of skilled on-site craft labour
- surplus factory space
- shortage of traditional materials
- need to replace war damaged properties
- encouragement, via increased spending allocations from central government to local authorities, to 'adopt new methods'.

During this period a total of 405,000 dwellings were constructed in the UK, mostly in two storeys but some three and four storey flat units. In 1953 the government withdrew its preferential grant for system building, and by 1959 a Building Research Establishment (BRE) report showed that there was little cost benefit when comparing non-traditional to traditional forms of construction.

Of this total of UK housing approximately 45% was designated 'defective'. But some other types not designated are now requiring renovation such as British Steel Construction (BISF) types and some Wimpey 'No fines'.

During this period there was an increase in the number of factories and offices constructed in steel or concrete frames. Generally these were no higher than four storeys. Concrete was usually in situ.

PHASE 3   1960 TO EARLY 1970s

In this period at least 750,000 and possibly upwards of 1,000,000 non-traditional dwellings were constructed in the UK. Also during this period an extensive school, university, college and hospital building programme was in progress. Many of these buildings were constructed in modern material combinations and/or system building techniques. The most common materials were concrete, followed by timber and then steel. In this latter material the Consortium of Local Authorities Special Project (CLASP) structures were erected. Although there have been some problems with this system, a constant programme of research and development has produced buildings of integrity and comfort – they could be considered as a success amongst hundreds of failures.

1.2   Checking cover to reinforcement in LPS block. *Solihull*

A constructional difference separated this phase from the other two phases, namely the extensive use of large panels as both structural and functional elements. These were used to build high rise, up to 22 storeys as well as medium to low rise. Reema houses are a low rise system, whilst Bison and Tracoba are examples of high rise LPS. The basic construction is shown in figure 1.3. This modern approach to housing people in high rise structures was new to the UK, though not in other countries such as the US. A few blocks of flats of five to six storeys had been built but were in the private sector and commonly in load-bearing brickwork. Tenements in Scotland have a long history and some of these reach to six storeys although, as in England and Wales, four storeys were common pre-1939. It is estimated that 140,000 new dwellings over five storeys were erected during this period. Not all were LPS but were in situ concrete frames with an external cladding of brick or precast concrete panels. These buildings are showing as many structural defects as LPS, and they are not confined to housing use. Many commercial buildings have this form of construction.

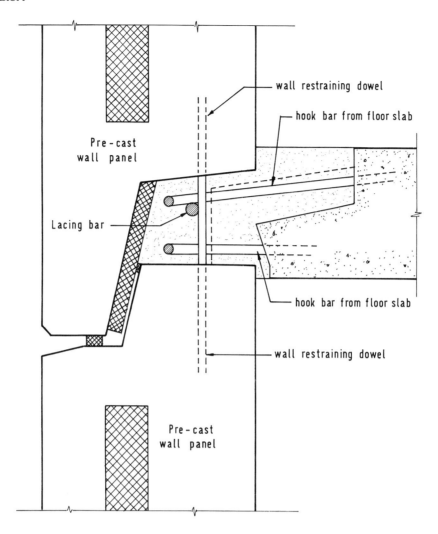

1.3 The connection between external wall and floor slabs for large panel system of precast concrete in high rise flats. *Tracoba system, Scotland*

The Federal Housing Administration later HUD (Housing and Urban Development Department) considered prefabrication in the 1930s. Some encouragement was given to factory produced units packaged to provide houses. These were in timber frame and low rise. Apartment blocks were being constructed in steel or in situ concrete frame with mainly brick claddings. Floor and roof were in situ concrete. There were no proprietary systems as were developed in Europe. But because the basic materials and form of construction are similar, common problems have developed, for example spalling concrete.

In the late 1960s HUD again reviewed the idea of mass producing housing. Under 'Operation Breakthrough' two types of proposals were called for. *Type A* – design, testing, prototype construction and evaluation of complete housing systems leading to large scale production. *Type B* – advanced research and development of ideas not yet ready for prototype construction or which might form part of the other total systems. In the event the major developments were based on low-rise timber framed. One scheme that was developed was for the Boston Redevelopment Authority. The four to twelve storey apartment blocks used precast prestressed concrete panels spanning 9.75 m. Brick was the main cladding material.

For further details and explanation of the scope and problems of non-traditional modern housing refer to the report produced by the Association of Metropolitan Authorities (1985) and RIDLEY (1988).

An example of the scale of the problem confronting public owners of housing is in Birmingham, UK, where there are 425 high rise blocks. Each of these now does or will require a degree of repair or refurbishment. This will be necessary not because of any major faults but through a natural deterioration coupled with higher (or different) social expectations which inevitably will occur. Therefore no building can be expected to stand for ever without some attention. All require maintenance and loving care and attention whatever the form of construction. Unfortunately it does seem that levels of maintenance and repair are not of the highest standards. This is true for both public and private owners although there does seem to be a wider scope of solutions in the private sector. What was not envisaged when designed and constructed was that modern methods of construction would prove to deteriorate at a faster rate than, say, traditional load bearing brick structures. As will be seen in chapter 2, there are problems peculiar to 'new methods', such as carbonation of concrete; poor steel reinforcement cover and poor workmanship during fabrication and site assembly. Couple these with relatively low levels of thermal insulation, ineffective weather protection and outdated facilities, such as bathroom and heating fittings, then this culminates in an unacceptable standard of accommodation. Where these defects occur at structural connections and in structural elements then the structural integrity can be affected. In other words the safety of the building may be in question. In the case of PRC houses a gradual deterioration, if not arrested, will lead to eventual collapse of a part or whole of the building. On high rise LPS structures strengthening may be required in order to resist the possibility of internal explosion removing structural panels, as seen at Ronan Point, London, in 1968. DAVENALL (1979); STROUD (1980). Cases of the external leaf of a cladding panel falling off have been recorded in Birmingham and Glasgow. This led to many owners carrying out programmes of inserting bolt fixings to panels to provide extra support.

CURRENT TECHNOLOGIES

Large panel systems using precast concrete panels are still commonly used throughout the world for the construction of dwellings. Also in situ concrete frames with a variety of claddings are utilised for commercial and housing accommodation. Unfortunately it is apparent that the mistakes of the past are still with us and will produce recurring problems in the future.

The Chicago Housing Authority has 40,000 dwelling units, 15,000 of these are in high rise concrete framed structures. Many of these are showing signs of concrete deterioration where exposed to the elements. Some are inadequately heated and suffer from condensation and lifts break down frequently. Window frames are leaking. There is now a major programme of upgrading, refurbishment and repair being enacted. This is in conjunction with alleviating social problems such as vandalism and muggings on the estates by increasing security. The building is being considered as a living unit in a social environment.

Central European countries have invested heavily in LPS buildings for housing. They provide virtually

1.4 An in situ concrete framed office building completed in the early 1980s now being repaired. Movement joints being renewed at junction of edge of concrete floor slab and brick cladding. *Offices, Birmingham*

all new housing accommodation and are constructed in large estates, which in turn create townships. The majority are high rise and in long slab blocks. It must be said that the standards of construction are certainly no better than those carried out in the 1960s. Indeed evidence shows that standards in prefabrication and assembly are much lower. Many panels suffer damage at their corners before and during assembly. Tolerances and fit are erratic. In situ site joints are not properly formed. Joint sealants are inadequate. Some attempts are being made to alleviate the problems. In Czechoslovakia simple forms of overcladding are being used to prevent weather ingress, as shown in figure 1.4 The problems was described as 'the gable end panels were full of holes and the wind blew through them'. On close inspection of this overcladding it was shown that steel screws had been used to secure the plastic sheeting to timber battens. These screws were rusting and some had already disintegrated. Chapter 9 raises the issue of the inadequacy of repairs and refurbishment technologies. In some cases the remedial measure has failed dismally and expensively.

1.5 Many countries are still using large panel concrete systems for housing. *Prague, Czechoslovakia*

THE REFURBISHMENT PROCESS

The following chapters present an introduction to the problems, methods and technologies that can be addressed in the refurbishment of modern buildings. It is not possible to give all the many solutions in either passing reference or detail – this would require a treatise of considerable length and illustration. The intention here is to describe the main stages in the process of bringing modern buildings to states of repair and comfort that meet the users' requirements. The bias is towards the technological issues although social, economic and legislative factors cannot be ignored. It is within the context of these factors that any technological solution will be made. Therefore this book will consider such factors where appropriate to their bearing on the refurbishment technologies. One of the fundamental factors affecting commercial building refurbishment is economic. Will the costs of refurbishment be covered by increases in rents? What

THE REFURBISHMENT PROCESS

standard of refurbishment will be required in order to attract tenants who will then pay a suitable rent? What further life is to be expected from the refurbished building? In housing, social considerations such as types of families, levels of income, expectations of comfort, relate to rent levels. Any improvements will be judged not only on economic grounds but also by the comfort and amenity provided. Forms of tenure are important. Many modern system built dwellings, in high and medium rise, have been sold to private developers. These companies have undertaken refurbishment programmes in order to offer flats for sale. The levels and types of refurbishment measures used range from minimal internal and external upgrading with no structural alterations to extensive external fabric and environmental upgrading as well as internal modernisation.

1.6 Plastic sheeting on timber battens providing weather protection owing to poor performance of concrete panels and joints. *Prague, Czechoslovakia*

1.7 In situ concrete frame and brick cladding showing signs of distress. *Chicago, USA*

15

In France, housing controlled by non-profit making associations, Societé Anonyne D'Habitations A Loyer Modre (HLMs), are undertaking general improvements to large panel systems building. Making a generalistic statement the French approach tends to show a much higher flair in architectural style when compared with UK refurbishment of similar properties. Extensive use of overcladding system (in brick, tile and sheet forms) is made. Extensions to increase usable areas are common. In order to meet changing family groupings, for example single people, one parent families, with no children and the elderly, structural alterations will be carried out. Holes for new doorways will be made in a precast concrete panels enabling floor layouts to be changed to provide a wider range of accommodation. These structural

1.8 Not only can the building be refurbished and overclad but further floor area can be provided by additions. *Grande Synthe, Dunkerque, France*

1.9 The LPS window panel will be removed to create an extension to the living room. *Dunkerque, France*

alterations are also carried out to external façades of LPS buildings.

So whilst this text is based on experience and practice in the UK, it also draws upon refurbishment knowledge in other countries to give some other ideas. The chapters follow the process of evolving a remedial solution to a defective modern building.

Chapter 2 describes the common structural and dilapidation problems found in these buildings. Mainly associated with chemical defects in concrete but influenced by design and workmanship.

In order to ascertain the problems detailed, an accurate diagnosis is necessary. Chapter 3 considers the methods of investigating low and high rise structures and puts forward a process for appraising the structure and fabric.

In considering options, a logical, well informed process (based on evidence and appraisal techniques) should be undertaken. It is at this stage that the technological appraisal is interwoven into the needs and constraints of the buildings' owners and users. It is here that sound technological solutions can be displaced in favour of poorer options owing to economic or other short term constraints.

The solution adopted must be clearly described and communicated to those who will be carrying out the work. The preparation and content of documents suitable for the range of remedial work is described in chapter 5.

The bottom line of virtually all decisions is: What does it cost? The factors affecting cost are described together with the particular points that are peculiar to modern building refurbishment.

Site organisation, chapter 7, looks at the problems for client, supervising officer and builder in undertaking the works. Dealing with people in place can take up to 80% of a site manager's time. Safety becomes an issue not only for site staff but for residents.

Chapter 8 presents a number of case studies showing the methods and solutions for modern building. As the majority are residential these are predominant, but some similar structures are used as commercial buildings and the principles are the same. Examples from France and the US are included.

Finally, chapter 9 considers the fundamental question of the value of refurbishment and provides a basis for assessing its worth. Also some further issues are raised regarding the performance of remedial works, some good news and some not so good. There is still much room for improvement and if this book has provided food for thought then it has contributed in a small way to the continuation of a better built environment.

CHAPTER TWO

PROBLEMS, PROBLEMS

Modern buildings suffer from a variety of physical problems. Most of these are connected with the structure and the external envelope. It was expected that the building's services, such as heating and ventilating systems, would become obsolescent at least once and possibly twice during the life of the structure. What was not envisaged is that the envelope and, in some cases, the structure itself would show signs of obsolescence after only 20 to 30 years.

Although not one building owner, designer or builder would categorically predict the structural life of a building, the expectation is that it would last for at least 60 years. The question of its economic and functional life needs to be considered as a separate issue. In the first place the basic structure and envelope should continue in good order, independent of these other factors. But comment must be made of these in order to clarify the meaning of the life of the structure in the context of rehabilitation.

The economic life is determined by:

- capital cost
- method of recovery of capital cost
- length of time for recovery of capital cost
- rate of return on rents, etc, where appropriate
- value of the building at any particular time
- value of land upon which the building is situated (as a single entity or in a group of buildings)
- cost of running and maintaining fabric and services.

The economic evaluation will also be influenced by social factors, such as:

- need for that particular building's function
- users' perception of the building
- land required for other purposes
- legislation/incentives by government to redevelop.

Blocks of flats, for example, in local authority (public) ownership with its capital cost borrowed over 60 years was initially expected to exist, in good state, for that period of time. Due to social factors, flats may be hard to let, running costs for tenants and landlord can be high. Defects are appearing in the structure which require some attention. The block is situated in an urban environment where land values are high and demand for accommodation of a high standard is strong. The social factor may determine the life of a building; rehabilitate; demolish and sell the land to a developer or rebuild to provide public rented accommodation.

A second example is of a shopping precinct near a town's main shopping area. Again, the structure and fabric are relatively sound and can be brought up to present-day standards. But the general layout of the shops does not provide suitable space for present-day retailing practices. The cost of the buildings would have been virtually recouped over its 25 years' life. Owing to its poor image and layout retailers find it hard to attract customers. In this case the precinct was demolished and a comprehensive redevelopment was undertaken which extended onto adjoining land. The commercial returns from redevelopment were seen to outweigh the structural life of the buildings.

The question of strategic options will be considered in further detail in chapter 4. The emphasis in this chapter is on the physical problems commonly found on modern buildings.

2.1 Large estate in city urban area undergoing refurbishment. *Doddington Estate, London*

2.2 Replacement of brick cladding and installation of cavity insulation. *Hospital, Leicester*

STRUCTURE

As the focus is on framed and panel buildings the main problems are with the concrete structures. Concrete is not the extraordinary material it was thought to be in the 1950s and 60s. It was considered to be strong, long lasting, inert and would require no maintenance. It is possible to produce concrete to meet these performance requirements and it is still very much a common material for tall framed structures. Far more is now known about the behaviour of concrete and it can be designed to meet specific requirements. Quality control during placing can ensure that the design criteria are met and that the concrete will perform to expectations. Unfortunately the level of design knowledge and site controls existing in the 50s and 60s was not high but, perhaps more significantly, what there was, was not applied. For example, the phenomenon of concrete shrinkage was known, but little or no allowance was made in the structure to accommodate this. An exaggerated example was found in a building constructed of large precast concrete panels. They were erected very soon after casting, and cured during the early years of the building's occupancy. The central heating systems dried out the panels. The problem manifested itself after about four or five years through the findings of the lift maintenance engineers. Since the early days of occupancy the lifts were constantly jamming. On most occasions the engineers had to adjust the runners relative to their fixing to the lift shaft walls. Eventually they used up all the adjustment and reported back to the owners. In total they had to reduce the lift run by 75 mm. In other words the 10 storey building had shrunk by 75 mm. This was attributed to concrete shrinkage.

DEFECTS

The problems in the structure could be:

- deflection in beams and floors due to weak design/unforeseen loading
- concrete creep
- carbonation of the concrete
- chloride attack
- corrosion of the reinforcement
- insufficient steel reinforcement placed
- inadequate/insufficient fixings between precast and in situ concrete components
- lack of sufficient load carrying packings between precast units
- corrosion of ties and fixings
- misalignment of precast concrete panels
- inadequate movement joints between claddings and structure
- poor sealants and joint design between precast panels and/or components
- poor quality control over in situ components and elements
- degradation of components due to lack of maintenance
- inadequate insulation leading to internal condensation
- surface finishes spalling or flaking
- distortion of panels.

A brief explanation of each defect is given here, but their consequences will be dealt with when addressing the methods of diagnosis and repair.

CONCRETE CREEP AND SHRINKAGE

Creep can be expressed as an increase in strain under a sustained stress. Obviously this is induced under conditions of load and can be still occurring up to 30 years, if not indefinately after initial manufacture of the concrete (NEVILLE 1983). This together with shrinkage produced by normal curing can cause deterioration of large structures, such as tall buildings. Creep can result in the gradual transfer of load from the concrete to the steel reinforcement in columns. Deflection in beams can be considerable if creep has not been taken into consideration in the structural design. There is some doubt as to whether creep movement can be recovered by reducing the stress.

A most important external factor affecting creep is the relative humidity of the air surrounding the concrete. In general terms the lower the relative humidity the higher the creep.

It is important to realise the difference between creep and shrinkage. Shrinkage is caused by the withdrawal of water from the concrete. This manifests itself in the reduction of overall dimensions of concrete elements as described earlier with the example of the 75 mm reduction in the height of a mult-storey block. Type of aggregate (its properties), water content of the concrete determine the extent of shrinkage, together with the curing regime. Another factor affecting shrinkage is carbonation.

CARBONATION

Carbonation is a result of the presence of CO_2 (carbon dioxide) in the air. The rate of carbonation increases with an increase in the concentration of CO_2 especially in concretes with high water/cement ratios. It is generally accepted that in the UK in the 1960s a significant proportion of concrete manufactured, especially on site, ultimately acquired a high water/cement ratio. Extra water was added to the mix in order to give it greater workability when placing in moulds and formwork. Therefore the design mix was altered. This factor may now be accelerating the rate of carbonation in certain structural elements. NEVILLE (1986) states that 'carbonation penetrates beyond the exposed surface of concrete only slowly'. In well-made concrete this may be so, but unfortunately experience has shown that carbonation can take place rapidly in poorly made concrete.

Carbonation is approximately proportional to the square root of time. This means a doubling from year one to year four; doubling again between four and ten years and a further probability of doubling up to 50 years.

Sulphate resisting cement used in the mix increases the depth of penetration by up to 50%; Portland blast furnace cement up to 200% more than with Portland cement. A guide to the normal rate of penetration of carbonation over time is (for average strength mix):

 5 mm depth within 1 year
10 mm depth within 2 years
15 mm depth within 4 years
20 mm depth within 10 years.

2.3 Spalling concrete due to reinforcement corrosion induced by carbonation. *University of Sussex*

The chemistry of carbonation is that the CO_2 in the air in the presence of moisture, reacts with hydrated cement minerals, CO_2 and moisture creating the carbonic acid agent. The $Ca(OH)_2$ carbonates to $CaCO_3$. Also other cement compounds can be decomposed, namely hydrated silica alumina with ferric oxide being produced. This results in increased strength and reduced permeability.

Carbonation in itself is not a problem, it is the consequences of this on reinforced concrete which has major repercussions. The alkalinity of the concrete is reduced from average values of 12 to 14 down to 8 or 9. It is the high Ph values which protect the steel reinforcement from corrosion. Therefore when carbonation penetrates through the rebar cover, and oxygen and moisture are present, corrosion will take place. The results of this are shown in figure 2.3.

When the concrete is subjected to alternating wetting and drying — a typical UK climate cycle — shrinkage due to carbonation (which occurs during the drying cycle), becomes progressively more apparent. This factor begets a thought: with the 'greenhouse effect' now becoming realised, that is increased levels of CO_2 and the likelihood of increased temperatures and rainfall, the rate of carbonation in concrete and its incidence could become greater.

The presence of cracks in concrete (due to other factors) can allow the ingress of CO_2, also oxygen and moisture, into the heart of the concrete. Carbonation and its consequences causes the greatest concern with the structural stability of some concrete buildings.

CHLORIDE ATTACK

In some instances additives were introduced into concrete to accelerate the curing. There were two reasons for this, one to achieve a quicker turn round of moulds or formwork, two to allow concreting to take place in cold weather. A constituent of these additives is calcium chloride $CaCl_2$. There is still some doubt as to the effect of calcium chloride on reinforced concrete, but evidence is emerging that it can have a disastrous effect by corroding the rebars. Chloride ions in concrete are not always distributed uniformly and their concentration has resulted in severe corrosion of rebars in structural members, especially where these are exposed to the weather. A number of PRC house types have been found to have $CaCl_2$ in their concrete elements, probably introduced in the factory to increase production turnover.

Also many prestressed concrete floor slabs have been used in many buildings. Evidence is again emerging that some will have been manufactured using $CaCl_2$ as an accelarator, then steamed cured. Its effect on prestressing wires, which are of small diameter when compared to normal rebars, is one of severe corrosion. On such a small diameter this corrision could be catastrophic as it will eat through them rapidly.

Since 1972 the UK Code of Practice has not permitted the use of calcium chloride in reinforced concrete or concrete containing embedded metal. Additionally the amount of chloride ion is limited when found naturally in the aggregates.

CORROSION OF THE REINFORCEMENT

Corrision of steel rebars is mainly due to oxygen and moisture. Any common steel will corrode slowly when exposed to the air with no protection. When ordinary carbon steel is used in concrete to give tensile strength, the actual concrete itself will provide protection. This is achieved in two ways; depth of cover from the atmosphere and alkalinity of the mix. Where depth of cover is small and alkalinity verging on a neutral Ph value then steel becomes vulnerable to corrosion. A very dense concrete can slow down this process.

Steel in corroding expands and can double its volume. Layers and lumps can flake off. As most reinforcement is embedded in concrete this expansion can push off the surrounding concrete.

This is seen in figure 2.3 where a vertical strip of concrete has been displaced along the line of the rebar. Once this has occurred the rebar is more exposed to oxygen, and moisture and corrosion is accelerated.

The prime cause of corrosion of steel reinforcement

2.4 Drilling for testing concrete: carbonation and dust samples for chlorides. *Solihull*

is carbonation together with a lack of adequate cover. It is obvious that a main steel rebar that has expanded and then lost its diameter due to flaking will not perform to its designed strength if the bars are left to corrode: Failure of the element will eventually occur.

INSUFFICIENT STEEL REINFORCEMENT

In some cases poor quality control on the site, and sometimes in the factory, has resulted in not all the steel rebars being placed in the mould. Therefore the concrete element will not perform to the design and, if overstressed, suffer from carbonation or other damage and will not have a sufficient factor of safety to cope. Premature failure could occur.

Although, thankfully, this defect is not too common, it must be borne in mind when carrying out investigations.

POOR FIXINGS BETWEEN STRUCTURAL ELEMENTS

This is far more common than the above defect. This defect is illustrated by the following findings:

- lack of stitch bars between precast units in in situ joints
- continuity rebars not passing through all connecting hoops
- connecting hoops and tie rods bent over where they did not line up
- hoops and tie rods burnt off where they did not line up
- ties in precast panels not pulled out to make connections with in situ elements
- where strengthening to LPS structures was carried out this was done inadequately, for example not grouting in the tie bolts
- inadequate levelling and securing of bolts between vertical precast panels.

SPREAD OF LOADINGS

Precast concrete panels are designed to distribute their load along the full length and area of their base. Bolts were used to level and position the panels. A concrete dry pack was inserted to transfer the loads from the bolts to the base. Two common faults can be found; one, the levelling bolts' nuts have not been slackened to transfer the load onto the dry pack; two, the dry pack was not inserted, or inserted only in short sections. On one building it was found that it was virtually impossible to be certain that the dry pack was properly inserted. The operative would push the material under the panel and it would slip away into a cavity behind. Pieces of cardboard, cigarette packets and other rubbish have been used to pack the panels. The outcome of this is that instead of the loads being uniformly distributed over the base area, they are concentrated onto the levelling bolts. The panels were designed to carry uniformly distributed loads (UDLs), not point loads. Undue stresses can be induced in the structure and its integrity can be rendered doubtful.

CORROSION OF TIES AND FIXINGS

Even where ties have been protected by using zinc coating there are numerous examples of corrosion occuring. Wall ties manufactured in the 1950s and 60s had a minimal coating of zinc via the dip process. Where these have been used on exposed walls high on multistorey buildings dramatic corrosion has occurred. Examples of complete loss of steel material have been found resulting in no physical tie between the two elements, namely the inner leaf and outer leaves of external claddings. Most instances have occurred with wire butterfly wall ties, but some corrosion is evident on strip steel ties.

Surface corrosion to unprotected steel bolts and fixings has been found, but owing to their relative large diameter it does not cause a problem. But where such steel fixings have been used to hold precast units at roof levels then any corrosion could worsen over the life of the building. Parapets and roof walls are particularly vulnerable to the effects of the weather.

Failure of a delta bronze tie used to connect the two leaves of a laminated precast concrete panel has been discovered. This was only realised when, at different locations, two panels failed by shedding their outer leaves. These fell to the ground from multistorey buildings but luckily did not cause any injuries. Investigation found that induced stresses had caused the ties to fracture and fail. This was confined to a particular type of delta bronze which had a limited manufacturing run. Further investigation revealed that the specified number of ties could not be relied upon to be found in place. This would cause further problems as those ties inserted would have to carry additional loads. The failure can occur when the stressed ties are in contact with water which acts as an electrolyte, drawing its chemicals such as chlorides from the concrete.

MISALIGNMENT OF PRECAST CONCRETE PANELS

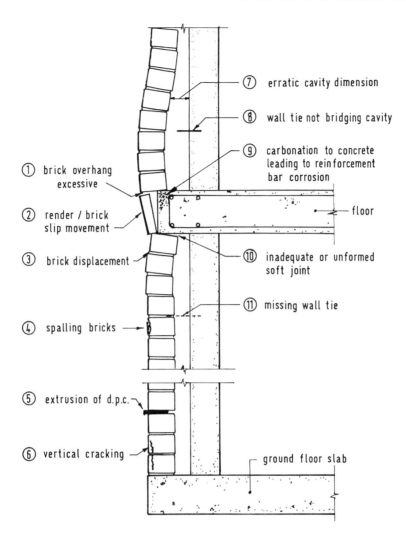

2.5 Defects to brick panels and structure

MISALIGNMENT OF PRECAST CONCRETE PANELS

Often the face of external precast concrete panels does not form flush surfaces, nor does it join squarely at a junction. In neither case is it a major problem, but it can lead to subsequent difficulties. The panels not flush with each other may have been erected that way due to:

- misaligned securing anchors
- varying overall thickness of panels produced during manufacture
- movement of whole panel after erection
- movement/delamination of outer face of composite panel.

A closer investigation will need to be made in order to ascertain the reason for the misalignment. Where only the joints are not matched then a poor weathertight joint may develop. Baffles and sealants may not protect the joint, and eventually rain, frost, etc, can penetrate causing degradation in the long term. In extreme cases water can penetrate into the building through these faulty joints.

2.6 Cleaning concrete parapet walls with pressure spray. *Bison System, Solihull*

INADEQUATE MOVEMENT JOINTS BETWEEN CLADDINGS AND STRUCTURE

This problem can occur on a variety of cladding materials and at a variety of positions relative to each other. Some examples are listed below giving their basic cause:

- inadequate soft joints between brick panels and concrete structures, seen in figure 2.5 (structural creep)
- lack of horizontal expansion joints on long runs of brickwork (thermal movement)
- differential movement between inner and outer surfaces of their claddings such as asbestos cement sheets (thermal)
- lack of movement joint between sheet cladding and retention frame (thermal structural).

The lack of adequate gaps to accommodate movement can result in undue stress in the materials causing bulging, cracking and spalling. Leaks can occur due to the breakdown of the sealants caused by this movement.

POOR JOINT DESIGN

In the 1950s and 60s the sealants used did not perform as expected. Many have failed by becoming either brittle (due to atmospheric contamination or to expansion/contraction cycles) or soft, and in both cases falling out of the joint.

Gaps between components could vary from 0 mm to 25 mm. A sealant in a designed gap of, say 12 mm will not cope with large dimensional variations. The sealant material could be unsuitable.

Where baffles, whether sheet metals or plastic sheet compounds, were inserted into grooves during erection this could be poorly executed. The baffles may not bridge the gap or be inserted for their full length.

In some cases where water has penetrated through open joints a few years after construction, the owners sealed the joints with mortars or mastic. This has not cured the problem as it is difficult to ascertain exactly where the water enters the joints. All that may be achieved is that water is trapped in the joints and will still eventually penetrate into the building. The evidence of water at a point in the building should not be taken as lining up with the point of entry: this may be many metres away, vertically or horizontally.

Poor attention in design to the way in which joints are made on site has caused some of the problems. The design solution was conceived with the expectation that ideal conditions would prevail. That is all the panels would align exactly, that a cold wind would not be blowing and that the operatives could handle the materials easily. In other words the design could not allow for any errors and would not tolerate even small deviations: the levels of accuracy demanded were too high using those materials and components at that time.

An example of insufficient thought being given to joint design is that between steel window frames and brick or concrete surrounds. In some cases the frames were placed next to brick or concrete and a mastic used to seal any gaps. This demanded high levels of dimensional accuracy; no distortion of the frames and good application of the mastic to the joint. A better solution was to place the metal frames in a timber sub frame. This enables a better fit between steel and wood. The timber could also be adapted to fit the actual opening sizes, with minimal use of mastic.

INADEQUATE SITE QUALITY CONTROL

This factor runs throughout most defects. Not that it is always the initiator of the problem, but lack of care during the site works can exacerbate an intrinsic defect. A few problems have already been described but specific mention is made of others as follows:

- inadequate mix control over in situ concrete elements
- omitting damp proof courses around openings
- not tightening connections or threading dowels through hoops
- damaging precast concrete panels during handling
- not placing reinforcement in the correct position
- not setting out the dimensions accurately on in situ elements
- ommitting wall ties.

The need for quality and conformance to specification (Quality Assurance) will be made when discussing methods of repair.

DEGRADATION OF COMPONENTS

Unfortunately many buildings have not been adequately maintained. This is due mainly to two factors: one, the attitude of 'if it is not causing any problems leave it alone'; two, lack of money to carry out planned maintenance. The window frames shown in figure 2.7 are an example of neglect. These are now totally decayed after 25 years of neglect and require replacement.

Services will also require renewal or upgrading. Buildings constructed before the early 60s will need rewiring as the systems will not meet present-day standards. Depending on the wiring materials used,

2.7 Poor maintenance has led to the premature deterioration of the softwood window frames. *Bison cross wall, Walsall*

some deterioration is likely. The insulation can become brittle and if disturbed may flake off.

In multistorey flats and heating systems, such as electric under floor, may not function efficiently and will need to be isolated. Other forms of heating systems will need to be appraised. Lifts are likely to be coming to the end of their useful life with excessive maintenance being experienced.

In office and commercial buildings the obsolescence of internal services could be the prompt to consider an investigation of the fabric and structure. Current demands for sophisticated communication and computer systems cannot always be fitted into existing envelopes. A complete reappraisal of the entire building will be necessary in order to ensure that the structure and fabric will provide the right performance standards to create a comfortable and economic internal environment.

The link of Legionnaires disease with water cooled air conditioning systems has prompted buildings' owners to look carefully at these systems. At best these will require regular and careful monitoring and maintenance. In the final analysis they might best be replaced. This in turn could initiate a closer look at all the building.

LOW STANDARDS OF INSULATION

A problem not fully anticipated in the construction of large panel structures was the possibility of internal condensation and/or interstitial condensation. Many forms of cladding were used. Curtain walls of glass and composite spandrel panels are prone to creating room condensation. Examples of this have been so bad as to cause condensed water to flow into lower floors. This was encountered on a recent refurbishment when new thin composite spandrel panels were inserted below steel window frames.

Sandwich panels of concrete and a fill of insulation have not performed well. On high relatively exposed corners of buildings excessive heat losses are occuring. This cooling of the external surfaces makes the panels act as a heat sink and can cause excessive room condensation. The panels themselves act as a cold bridge.

SURFACE FINISH FAILURE

A common failure is spalling of mosaic finishes. In bad cases debris can rain down and safety canopies have to be erected, as shown in figure 2.8. The dislodgment is caused by:

- structural movement setting up excessive stresses
- lack of bond between mosaic tiles (or sheets) and the background
- steel reinforcement corrosion (initiated by carbonation) pushing off mosaic
- freeze/thaw cycles causing the mosaic to become loose.

Concrete itself can spall, whether as the surface to a panel or the edge of a beam or column. Usually this is caused by rebar corrosion with its subsequent expansion.

Where brick panels have been excessively stressed, spalling can occur. This can be accelerated by the freeze/thaw cycle.

Some self coloured window spandrel panels can fade or stain. This is not a safety problem but can make a building appear dilapidated.

DISTORTION OF PANELS

Concrete panels can bend or twist due to a combination of factors:

- inadequate gaps for movement (no inbuilt tolerances in the joints)
- finished in dark colours
- placed on elevations exposed to the sun (south facing in the northern hemisphere).

Such an occurance was first noticed internally where a vertical gap opened between a partition wall running at right angles to the external panel. The inhabitant could see through to the adjoining room. The external panel was bowing from the face of the building creating this gap.

PROBLEMS WITH OTHER BUILDING STRUCTURES

2.8 Protecting access from falling spalling concrete and mosaic. *Erdington, Birmingham*

Steel framed buildings

Steel framed buildings are not exhibiting the range of failures associated with concrete structures but there are a few examples of defects in steel framed houses. As would be expected, the defects are primarily ones relating to corrosion. The most vulnerable point is where the columns are connected to the foundations. Rising damp, an unventilated space and unprotected steel can produce corrosion.

Steel sheet claddings, fascias and soffits have also been used on two storey houses. Where these have been well maintained few problems occur. The points at which corrosion is apparent is at the corners of roof soffit and verge. Rain water can penetrate here and settle above the soffit. Eventually the corrosion eats through to become visible externally.

There have been cases of profiled sheet claddings requiring continuous repainting as the finish flakes off quickly. It is not clear what causes this, but it is thought to be a condition of the sheet's metal surface originating in its manufacture.

Timber framed structures

Problems are arising in timber framed structures. Two and three storey school buildings constructed in prefabricated timber panels are suffering from wet rot. Rain water has penetrated into wall cavities and has not escaped. Over time, 15 to 20 years, the timber has deteriorated. The prime fault lies in the inability of the external cladding to prevent weather penetration.

The inadequacies of brick cladding to timber framed houses has also resulted in major defects. An estate of 75 houses constructed in 1976 has had to

2.9 Rainwater penetration through the brick cladding on poorly protected ply sheeting of timber frame panels. *Nuneaton, Warwickshire*

be completely reclad. The problem showed itself when the top courses of brick to a gable end fell off. This was caused by differential movement between the timber panel frame and the external leaf of brickwork. Inspection showed that inadequate movement gaps had been left to accommodate movement. Water had penetrated to the wood ply sheet and caused it to distort and rot. The brick/panel cavity was found to vary in dimension, in some cases the surfaces touching. The sheathing membrane to the ply did not cover all the surfaces. On one block of houses where there should have been 1350 wall ties only 300 were found in place. Plasterboard nails had been used to secure these to the ply sheet rather than screw nails to the timber stud frames – there was little or no structural integrity between the outer brick cladding and the timber frame. Insufficient cavity barriers had been inserted to prevent possible fire spread. Insulation and the vapour barrier were missing in a number of areas. Most of these defects were attributable to poor workmanship during the site operations.

The use of untreated timber in the frames until the late 1970s must cause disquiet regarding the durability of timber framed structures. The weakness appears to be in the weather proofing of the claddings. If these are sound, then water cannot penetrate to the timber members and start the decay process. The main problem with the claddings can be summarised as:

- inadequate joint tolerances to take up any differential movement between frame and cladding
- poor sealing of joints between cladding components
- irregular cavity dimensions creating bridges for water.

The emphasis of this book is on the physical problems and their remedies on modern buildings, but these must be seen in the context of other influencing factors, especially in the case of housing. The physical problems can be initiated or aggravated by the buildings' users (or third party abusers). Without an understanding of how people use buildings; what their circumstances may be; how they view the building (investment, workplace, home, leisure activity environment, alien atmosphere); why they are there; what do they want from it; how will they look after it; any remedial measures are likely to fail. Indeed in the case of housing the PRC concrete estates are themselves an example of little thought being devoted in their design to the expectations and requirements of people under tenancies. When upgrading or repairing buildings it is important to be fully cogniscent of the needs of the buildings' users. This applies whether they are renting or buying. Therefore before moving on to consider further aspects of diagnosis and repair some discussion should be devoted to social issues influencing repair and rehabilitation.

SOCIAL INFLUENCES ON BUILDING DEFECTS

This discussion is centred on problems associated with housing, particularly that in the public sector controlled by local government. There are also repercussions for private owners of PRC type houses. Finally, issues will be raised relating to the incidence of defects and patterns of use in commercial/factory/-public buildings.

The clearest analysis put forward so far regarding the correlation between design of buildings and the behaviour of users is by COLEMAN (1985). Her basic premise is that there is a significant correlation between the design of the housing and the incidence of social misbehaviour. The study considered housing estates constructed in the 1960s. These were predominantly constructed in concrete, with various forms of layout shapes and height of building. For example, samples of high rise point blocks; point blocks and slab blocks on raised columns (open ground floors) and slab blocks with balcony access at various levels and linking separate blocks. A total of 4,099 blocks were investigated, centred on two London boroughs and an estate at Oxford. The evidence of social malaise was based on ascertaining the levels of visible traces of 'misbehaviour'. These were: litter, graffiti, vandal damage and excrement. The frequency of their occurance was measured and plotted against the particular blocks according to the following design variables:

size
dwellings per block
dwellings per entrance
storeys per block
storeys per dwelling

circulation
overhead walkways
interconnecting exits
vertical routes (lifts and staircases)
corridor type

entrance characteristics
entrance position
entrance type
blocks raised above stilts
blocks raised above garages

features of the grounds
spatial organisation
blocks in the site
access points
play areas

Some of these variables had been identified by NEWMAN (1972) when considering the concept of defensible space in housing communities. The general conclusions reached are given as the influence that design has on the incidence of litter, graffiti, excrement and vandalism and the partially assessed factor of children having to be placed in care (based on information from one local government authority). The list is in a hierarchy from strongest design influence to weakest design and the numbers are mean percentage ranges for the test measures (litter, etc):

Dwellings per entrance	57.7
Dwellings per block	46.7
Storeys per block	41.1
Overhead walkways	32.6
Spatial organisation	31.2
Spatial organisation incl houses	48.2

(incl semi-private space of gardens)

2.10 Mesh and security grilles on access balcony. *Chicago, USA*

Vertical routes	26.7
Access points	24.6
Interconnection exits	22.4
Corridor type	20.6
Blocks per site (excluding pollution)	18.3
Storeys per dwelling	13.8
Blocks per site	11.7
Entrance type	9.8
Entrance position	8.5
Play areas	7.1
Stilts and garages	6.6

Therefore the strongest design factors that influence social behaviour are dwellings per entrance (how many are served by a single entrance – for example in a point block 100 flats may be served by just two main entrances); dwellings per block (how many separate units there are in block – high rise or medium rise); storeys per block (the greater number of storeys the greater the evidence of social behaviour); spatial organisation (this means the general layout of the estate, the amount of common and shared ground, who owns what, the relative position of one block to another, the possibility of surveillance of one block over another to give 'defensible space'). These design factors have been classed as design disadvantages by COLEMAN. In other words, where these occur they are seen as a negative effect on people's behaviour. This is reinforcement by comparing crime rates and incidence with the design disadvantagement score (expressed as a number to give comparison) and where this is high, so is the incidence of crime such as burglary, theft, criminal damage, bodily harm, sexual assaults and robbery.

Although the major concern must be with crime against people, as technologists it is worrying to see that damage is correlated with design disadvantage. The fabric and the structure of the building can be affected by intentional damage. Even the fouling by urine can have a detrimental effect (let alone the hygiene concerns) on the building fabric. Damage such as broken doors, smashed spandrel panels,

2.11 Impersonal common areas with deck access. *Portsmouth*

broken windows, stripped sealants, can initiate or decelerate any defect.

There are reservations about the import of these findings on the rehabilitation strategies required on modern housing estates. There is little doubt that there is a design influence on the social problems on such estates, but to what extent this determines behaviour is disputable. What is not in doubt is that if these buildings are not looked after then defects will develop.

Another aspect of user behaviour is the manner in which the internal environment is controlled, mainly with respect to heating. Due to economic reasons occupiers may not heat all rooms and/or heat rooms intermittantly. The use of washing machines (in domestic accommodation), greater frequency in bathing, can generate high levels of humidity. This coupled with low or irregular space heating can aggravate the rates of condensation. This in turn can lead to mould growth, timber rot, plaster falling off and so on. This can also apply to commercial or factory buildings where care is not taken over the control of the internal environment. In traditional brick built structures these problems are not so apparent as the materials more readily absorb excess moisture and can cope with irregular heating patterns. Relatively thin concrete cladding panels, metal windows and exposed external surfaces can make the internal environment difficult and expensive to sustain at comfortable levels. In a large block of dwellings it is likely that few flats will be well heated; some partially heated; single rooms only heated or no heating at all. This means that heat flow, and thermal change producing material expansion/contraction can create widely different performance characteristics in the structure. This will have some influence on the rate of deterioration.

The use and upkeep of all buildings will affect the levels of maintenance and repairs. A poorly maintained building is more likely to be abused by its users than a well maintained building.

The issues of social behaviour, levels of maintenance and poor internal environmental control will be raised again when options for repairs and repair methods are considered. These factors do influence strategies and solutions.

Summary

From the list of defects presented it is evident that the range is wide and that their effect can be catastrophic. Most of the defects are associated with concrete. Until recently this material was thought to have virtually an unlimited life; this has proved to be false. Concrete can last a long time but it does have inherent weaknesses and these can drastically reduce the design life. Fixings into concrete and between components are also vulnerable to failure, depending on their material and usage. Problems not confined to concrete structures, timber and steel framed buildings show various forms of degradation.

The influence of poor workmanship and design can exercerbate the problems. In some cases these may initiate the defect. The incidence of faults on traditional housing is apportioned 50% design, 40% workmanship, 10% materials (BRE 1984). There is no reason to suppose these proportions may be any different on modern framed structures. They were built by the same professional teams based on the same technological and educational backgrounds. If after 150 years of building in load bearing brickwork defects still prevail, what hope was there for innovative and untested systems?

The effect of some of these systems in the housing sector has led COLEMAN to postulate that their overall architectural form and layout has produced aspects of social malaise. This in turn has led to an exaggeration of the defects to this type of building. User behaviour can affect all building types and physical problems should be placed in the context of patterns of use.

CHAPTER THREE

DIAGNOSIS

It is essential to ascertain as clearly as possible the exact nature and extent of the defects in a building. Unfortunately experience has shown this is very difficult to achieve. The main problem is in determining the extent of the defects. The levels of confidence in this aspect will vary from building type to building type. The more buildings of a similar construction are surveyed the greater the likelihood of being able to predict the extent of the defects. Even comprehensive surveys do not present an accurate picture because many defects only become apparent when the building is opened up. Unless all the building is thoroughly investigated no definitive measurement can be made of the extent. It is possible to identify the nature and range of defects by carrying out sample checks. For instance, an opening up of in situ stitch joints to large panel construction, the removal of a few cladding panels to look into the inner recesses and the taking down of a brick panel at ground level to ascertain the state of wall ties, cavity, etc. Material samples can be taken and analysed in a laboratory.

A systemic approach must be made to the whole process of diagnosing the buildings problems. Overall, a number of stages should be undertaken. These are shown in figure 3.1, MOTTERSHAW 1988. A brief description of these stages is now presented.

The process should start with the good practice of regular inspection, or inspection following on from a defect being logged into the building owners' system for maintenance. There should be an inbuilt referral 'alarm' warning for any unusual or known to be serious defects to be highlighted quickly. The responsibility for initial awareness of the problems can only come from the building owner, but after this the professional advice should be based on the owner's requirements.

The owner (client when seeking external consultants' advice) should set down the brief for investigation in writing. An effective programme of investigation should always allow for reassessment and further consultations, developing the brief where necessary. This is the prime reason for adopting a phased approach. The other reasons are to enable budgetary control to be maintained by the owner/client and realistic dates to be set for reporting. When the brief has been defined the investigation should be directed towards the solution of the problems requested; the exception to this rule is where safety is concerned. Where the safety of occupants or users of a building and the surrounding area is in question the owner/client should be notified and action taken to eliminate the safety hazard and rectify the defect.

The appraisal may include any or all of the following:

- overall stability/strength/stiffness
- movement control; materials/loading
- weather tightness
- durability; life expectancy
- fire resistance
- safety.

The defects discovered may be categorised under the following:

- defects in design
- defects in construction
- deterioration since construction
- accidental/deliberate damage
- changes of ground/external environment
- change of use/increased loading
- changes in pattern of use; internal environment.

DIAGNOSIS

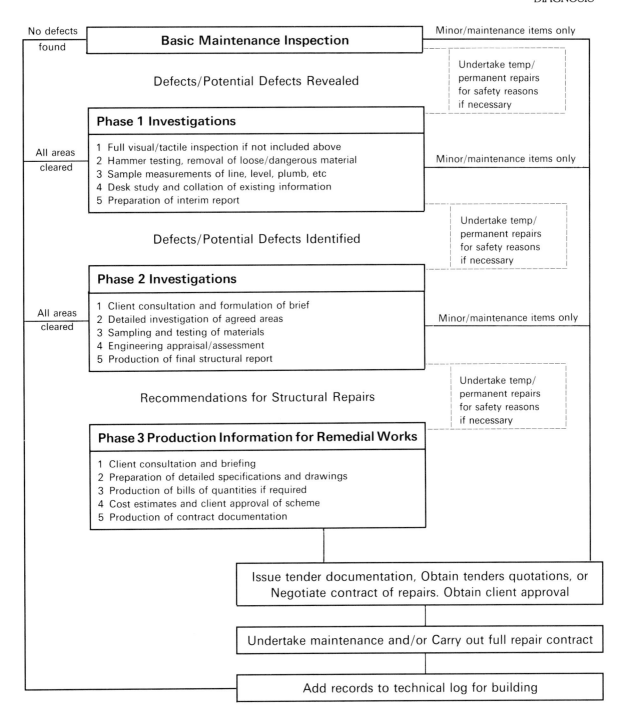

3.1 Programme flow chart for determining remedial work. *Mottershaw, TERN Project 1988*

PHASE 1   INVESTIGATIONS

The purpose of these investigations is to obtain a preliminary description of the building from a study of its documented history and a close look at the structure and fabric. Sample measurements should be taken where problems are obvious.

The desk study should be centred on the existing information regarding design and construction. The sources can be:

- original design information including designed floor loading capacity and geological investigations
- original drawings of the structure including general arrangements, elevations, reinforcement details and details of connections
- any site records of construction, Clerk of Works' reports, etc
- details of any modifications, strengthening, additions or alterations to the structure since construction
- the maintenance history of the building, including results of previous maintenance inspections
- any previous reports, surveys, investigations.

All too often, the above information has not been available in whole or in part. No systematic 'log book' has been kept of many buildings, whether in the private or public sector. Original design details may have been lost; the original builder not now in business and records destroyed; the reluctance of designers to part with information without payment of a fee. The lesson to be learnt is that a comprehensive log book should be kept to avoid unnecessary and expensive investigations later in the building's life.

The phase 1 investigations should be more than a look from a distance. Low rise structures can be viewed from the ground, but to obtain a better understanding a literal 'hands on' approach should be undertaken. Hence the reference under 2 in the flow chart the removal of loose material. This applies equally to multistorey buildings. Certainly binoculars do not give any real indication of the condition.

Based on the preliminary investigation an interim report should be prepared.

PHASE 2   INVESTIGATIONS

If after phase 1 investigations, there are defects or potential defects identified then a further detailed investigation will be necessary. Again a clear brief should be developed by the owner/client so that the limit of the appraisal is known. This will culminate in the preparation of the final structural report. This should indicate options for remedial work or recommend other courses of action.

PHASE 3   PRODUCTION INFORMATION FOR REMEDIAL WORKS

This will be prepared after a full appraisal of all options has been carried out. This will be discussed in chapters 4 and 5.

METHODS OF INVESTIGATION

The methodology of investigation is based on a systematic set of procedures. This will apply to high or low rise buildings. The main difference between the two will be of obtaining access and a more extensive appraisal of the high rise structures.

HIGH RISE BUILDINGS

A checklist of an inspection of a façade to a high rise building is given in figure 3.2

Generally it is best to work from the top starting with the roof. A grid system for the inspection will confine the inspection areas into manageable sizes and provide an identification. The width of each zone is determined by the method of access.

If the building is an LPS then it will also be necessary to check the internal panels. The structural integrity and condition of typical joints need to be assessed. At least three to four in different locations to give a fair indication to their state. Figure 3.3 provides a checklist for this internal inspection.

The inspection of internal panels requires the opening up of joints in addition to testing their general state. The checklist below lists the items needed to ascertain the general condition of the panels.

3.2   Checklist for inspection

Information Item	Reason for examination or objective	Technique	Comments
Joints	Misaligned	Visual	
	Movement	Visual – tell tale	
	Settlement gaskets	Visual – tell tale	
	Sealant	Visual	
	Gap width	Measurement	
Facing	Check for fixity	Low	
Concrete	Carbonation	Phenolphthalein test	
	Chlorides	Take dust sample and test	
		Take dust sample and test	
	Cement content	Tap with hammer	
	Spalling		
Reinforcement position		Cover meter	
Cladding panel Tie position and type	Presence and distribution	Cover meter Extract one from end panel for ascertaining material and state	
Alignment of panels		Visual and measuring devices	
Extent of cracking		Visual and measuring record by diagram and photograph Use BRE classification systems for crack description	
Windows		Visual Damp probe in timber	
Balconies Railways Soffits Parapets		Visual and tactile tap with hammer	
General		Visual and tap with hammer	
Connections		Opening up one on each elevation to check inside	
Copings		Visual and tap with hammer Cover meter for connections	

Item	Objective	Technique	Comments
Concrete	Carbonation chloride	• Carbonation • phenolphthalein test • dust sample and test for cement content chloride ions and cement	
	Cover to reinforcement	• Cover meter	
	Presence and form of cracking	• It may be necessary to take off plaster finish or lining to expose concrete surface	
	General conditon	• Check for moisture content with meter • look for mould growth, etc • inquire about users' room temperatures and comments on draughts, etc	

The structural integrity and condition of the typical joints need to be assessed. This will require opening up selected joints in at least two flats in different areas of the building.

This frequency should be adopted for most high rise structures, whether in steel frame or in situ concrete. By doing this a sequence of 'snap shots' of the condition can be recorded which will allow a complete picture to be built up of the building's behaviour. For example, a thin crack found in the first investigation does not show any signs of movement after four years. This knowledge will indicate that there has been no further movement since the crack first opened – therefore might it have been immediately after construction, or was there some settlement at another time which has stabilised? The answers may well be unclear, but at least the record shows no further movement. This would not have been known without regular inspection.

Figure 3.5 shows the positions for inspection and possible problems for a high rise large panel structure.

A number of examples centred on structural systems will show the results in practice of investigations.

3.3 Checklist – condition of panel

THE BISON WALLFRAME SYSTEM

Brief description

The internal panels are solid reinforced concrete units.

The external panels are a composite manufacture and the outer cladding skins, with either mosaic tiles or exposed aggregate finish, are 75 to 80 mm thick.

They are secured to the inner load bearing 100 to 150 mm thick leaf by 'goal post' type ties. These are approximately 150 mm long and 6 mm diameter. The material from which these ties are made varies according to the time and period of manufacture.

Between the inner and outer leaf is a 25 mm layer of polystyrene insulation with a damp proof membrane and this is penetrated by the ties.

The panels were cast in horizontal moulds either 'face up' or 'face down'. The joints used were in situ

THE BISON WALLFRAME SYSTEM

Item	Objective	Technique	Comments
Bearing of floors on panels			
Type of connection			
Condition of connection			
Embedment and bond of connection			
Presence and alignment of connection			
Presence and condition of full/dry pack under panels			
Condition of in situ concrete joint filler			
Type of condition of packing to top of panels under floor slab			

concrete surrounding dowels and hooked bars. Figure 3.6 shows a typical detail of joint between external panel and floor slab. Figure 3.7 shows a typical inner panel joint between floor slabs and upper panels.

Checklist for internal inspection of Bison Wallframe system

The process is as follows:

- select one wall of one room within a dwelling
- remove any furniture or fittings on or adjacent to the wall
- erect curtain from ceiling to floor to prevent spread of dust, etc

3.4   Checklist – joint information

- remove skirting board to expose dry pack
- check dry pack for soundness
- remove area of insulation at floor level adjacent to in situ wall joint
- take dust sample of in situ joint
- also take dust sample from pc panel adjacent to the joint
- inspect internal wall and take dust sample
- remove area of insulation at ceiling level
- open up wall to expose mechanical joint for inspection

DIAGNOSIS

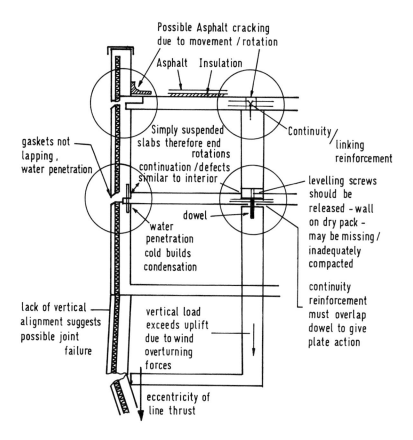

- in top floor flats, expose hairpin tie restraining coping element.

General findings

Findings have varied greatly from block to block. Some panels on blocks have been found to be in very poor condition whilst others are sound.

Chemical attack

Carbonation depths vary from 2 mm to 36 mm in extreme isolated cases. The average running to 12 mm.

Generally high levels of chloride ions are present in a majority of panels. This, together with carbonation will put a considerable risk on the reinforcement bars in the panels.

Mosaics

External mosaics are not sound. When those spall and expose the underlying concrete the rate of

3.5 Points to investigate on high rise large panel structures

deterioration will increase. On some blocks it is likely that the mosaic is in very poor condition. Remember, spalling can injure building occupiers and passers-by.

Joints

The joints between panels on the facade are usually covered in mastic which prevents inspection of the gap. Although in many cases this gap is non-existent.

Stress can occur between panels owing to thermal and moisture movement if they are in contact with each other.

Panel fixings

Where early patch repairs have been carried out these are now failing.

Perhaps the greatest factor of uncertainty is the outer leaf to the inner leaf of external panels.

THE BISON WALLFRAME SYSTEM

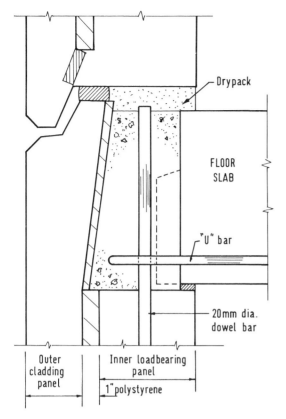

3.6 Joint between external panel and floor slab. *Bison wallframe system*

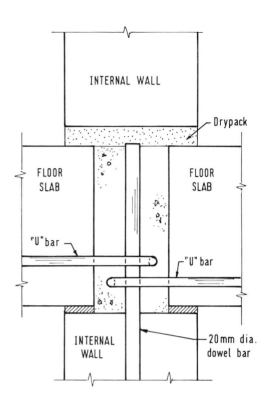

3.7 Internal joint between floor slab and upper panel. *Bison wallframe system*

A number of these outer leaves have slipped or fallen off. If not already done, it is strongly advised that these panels should be rebolted immediately.

Ties

It has been found that the 'goal-post' ties cannot be relied upon to hold the outer leaf to the inner leaf. Investigations have shown that the ties have only been pushed partially into hardened concrete causing voids to form.

These voids reduced the area of bond between concrete and the ties. In some cases the bond length had been reduced due to insufficient depth of penetration into the cladding leaf.

Less than the specified number of ties had been inserted and some were positioned incorrectly.

Tests have also shown that 'sticking' to polystyrene is adequate.

Stress corrosion cracking had been suggested as a predominant factor in the failure of ties made from delta bronze. These ties were used in blocks constructed from 1960 to 1968.

Generally, the mechanical connections to all the floor joints are sound, but it is likely that the 'U' bars could be missing.

Summary

The inspection of Bison wallframe buildings follows the general process described earlier.

The particular points to note are:

- spalling mosaic facing
- internal ties to sandwich panels
- progressive deterioration of reinforcement due to carbonation and chloride ion content.

FRAMED STEEL CLAD TWO-STOREY HOUSES

Examples of systems using this form of construction are:

Atholl
Birmingham Corporation (House No. 2)
BISF (first floor)
Braithwaite
Cowieson
Hawthorne Leslie
Howard
Keyhouse Unibuilt
Riley (first floor)
Weir

Characteristics

This category may include timber framed steel clad construction, but timber frames used in this form are comparatively rare.

- It is more normal to find steel framed systems clad with other materials. As with previous categories, the type of frame used varies widely among different systems.
- The outer sheet cladding is impermeable to driving rain, eg asbestos or sheet steel.
- These will probably be a traditional concrete strip footing. A good example of this occurs with the Atholl system. As usual there is some variation within systems.

Problems

Problems that occur with this type of construction:

- the sheet cladding will normally be impermeable to moisture penetration – particularly driving rain. But condensation may occur on the back of sheet steel cladding
- if thermal insulation is used in the cavity, condensation risk increases because of the lower temperature of the back of the sheet steel cladding. Lack of ventilation in the cavity increases the risk
- the effective vapour barrier may be in the cold side of the insulation
- corrosion of the primary steel frame, secondary frame and metal connection of the sheet cladding becomes more likely.

For a detailed analysis of many types of steel framed house, reference should be made to 'Low rise steel and timber', S McCabe TERN Project, Birmingham Polytechnic, 1988. This also describes the common problems associated with these structures and how to investigate them. The Building Research Establishment has produced a series of books on these types of houses.

REEMA HOLLOW PANEL HOUSE

General description

The basic structural form of a Reema house is:

- wide storey height precast lightly reinforced concrete panels
- these panels are hollow double skinned having channel-shaped rebates cast in their upper and vertical edges
- into these rebates an in situ concrete ring beam at first floor and eaves level and columns was cast
- a mastic joint was placed between the panels.

There are minor variations in the form of construction and also variations in the condition of the houses from one part of the country to another.

Generally, the further away from the place of manufacture, the greater the problems. It's assumed that as the distance between factory and the site increased, the level of control over site activities decreased. This is borne out by evidence that problems in connections and the in situ elements can be put down to poor site quality control. This type of information – location of manufacturing factory – should be gathered during the desk-top survey.

Investigation

The investigation procedures and the problems to look for are described in figure 3.8. Inspection points as a reference are:

1 Eaves level in situ concrete ring beam
Reach this via the roof space. This will exhibit problems and the concrete is generally quite poor. It is honeycombed and weak, but another beam can be rich in cement and strong. Test for carbonation and chlorides.

3.8 Opening up first floor to inspect ring beam.
Reema House, Bristol

2 First floor in situ concrete ring beam

A section of the wall will need to be broken open to gain access to this beam. This is done internally at the first floor level. By opening here it will also be possible to inspect the concrete floor beams (if present). Again, this ring beam is likely to be very poorly made. Remember that the beam literally ties the building together, so it is important that its correct condition is found. Some reinforcement steel may also be present.

3 Concrete floor beams

Generally the bearing of these beams is satisfactory, but the major problem is the possibility of a massive chloride content. Include a high proportion of these beams in the inspection. Take floor boards up at random points over the floor area and drill out samples and analyse the dust for chloride content. It is common to find medium and high values in a significant proportion of these beams. For example, in manufacturing during cold weather calcium chloride was shovelled in to the concrete mix to speed up curing. This enabled moulds to be stripped more quickly ready for re-use.

Some end to end cracking has been found on these floor beams and also spalling has been discovered due to rusting reinforcement. The beams are overdesigned and sudden collapse is not likely, but carry out a thorough inspection. It has been found that a beam can have a high chloride level at one end and a low at the other.

4 Bottom of panel

This requires inspection to assess the bearing of the panels on the floor slab. Take samples from here to test for chloride and also carry out a phenolphthalein test for carbonation.

5 In situ concrete floor

Take core samples from the floor to ascertain:

- possible sulphate content
- possible chloride content
- thickness of slab

DIAGNOSIS

- inclusion of any mesh reinforcement
- carbonation which may affect any reinforcement.

6 Foundation

Dig a trial hole by the side of the foundation and find the depth of the foundation. It may be that there are variations on the types of foundations used and the soil should be analysed to find:

- its bearing capacity
- presence of any sulphates.

7 Window reveals

There is reinforcement in the panels around window openings. Check for any cracks in this region. Also check the reinforcement cover using a cover meter.

8 Face of panels

Many houses have been overcoated with masonry paint. Unless this is stripped off it may be difficult to detect hairline cracks. If houses are well protected it is worth establishing whether such protection has been applied since construction, and exactly when. Is it very recent?

In the majority of houses there is no mesh in the main face area of the panel. As previously noted, the main steel reinforcement will be found underneath the sills of windows and to the jambs. Also likely to be located are the steel lifting hooks.

Panels do move due to thermal variations. Check to ensure that there is no excessive displacement.

It is also common to find a 200 mm wide length or 3 mm diameter 100 mm square mesh as a vertical strip at the extremity of each panel. This was embedded in both inner and outer skins.

Apparently this was inserted to mitigate any handling damage. If damage occurred the skins could be easily repaired as the vulnerable corners would be attached via the mesh strip. Repairs could be carried out in place on site.

The joints between the panels are likely to have been repointed using a high quality mastic and this should prevent any ingress of water.

9 Internal fibre board

Internally no damage whatsoever to party walls has yet been discovered. Carry out checks on all walls internally. Note that the plasterboard or fibreboard linings to the panels can and do come loose. Generally this is found to be a breakdown of the bond between the fibreboard and the concrete and is easily repairable. These areas of hollowness have been interpreted by occupiers to imply serious defects.

10 In situ corner columns

As with the case of the in situ ring beam, the in situ columns may well have been poorly constructed. The concrete can be honeycombed. Corner columns are most vulnerable to deterioration due to exposure. It is possible to repair these columns using standard concrete repair techniques.

General points

The following general points should be investigated:

- is there any differential settlement across the house or pair of houses?
- look at the condition of the chimney stack. What is it made of?
- check the roof coverings and the carcass
- make sure the rainwater goods are in the normal position as a standard unit
- are there any other extensions or additions?
- are there any differences in levels across the ground and first floors?
- determine the materials used for the internal partition walls. If these are concrete then test for steel reinforcement, depth of cover and the usual chemical analysis
- check to see if there are any other additions.

TIMBER FRAMED HOUSES

The inspection

Methodological inspection should enable you to build up a picture which will confirm whether or not the house is of timber frame.

Appropriate procedures are described below, followed by check lists.

External inspection

Windows in timber and brick houses are often set back in the wall, fixed to the inner leaf, but this is not always the case.

Between the window frame and the brickwork, at sill level, there should be a compressible layer, perhaps 5 to 10 mm thick, depending on how much compression has taken place. This may have been painted over or otherwise disguised. However, beneath the timber sill and above the brickwork, there should be something slightly soft and spongy (again, partly depending on degree of compression).

Some designs, such as Crest homes, were detailed in such a way that a flexible layer was not needed – so even if there is no flexible layer, particularly in older properties, it could still be a timber and brick house.

Internal inspection

Walls
Tapping the plaster work of external walls generally indicates whether or not the wall is dry lined (ie lined with plaster board). If the wall sounds slightly 'hollow' at most points, then it is possibly dry lined. Timber and brick construction uses this type of internal finish although it is also used in many other forms of house construction.

In timber frame housing the area above window openings should sound 'solid'. Whereas dry lined house walls will still sound 'hollow'. This is where a timber lintel will have been used. Using a sharp bradawl you will easily penetrate the plasterboard (about 12 mm thick) and will then feel it entering wood if the building has a timber lintel. A position that is behind a curtain or pelmet should be chosen so that decorations are not spoiled.

However, before making holes everywhere, a visit to the roof space could give more reliable and less disruptive clues.

First floor details
Sometimes the first floor may have an easily removable floor board as a service trap. Look along the length of the floor joists to where the joist is supported on an external wall. If it appears to be on a timber head plate, the house is almost certainly timber framed. If blockwork is visible, it is almost certainly not timber and brick.

3.9 Erratic tiles to window sills showing differential movement between timber frame and brick cladding. *Nuneaton, Warwickshire*

Loft space

If the gable end is in blockwork the building is almost certainly not timber and brick. A full gable end triangle, clad in plywood (or fibreboard) almost certainly indicates timber frame construction.

Semi-detached and terraced properties will have plasterboard (25 mm thick, comprising two 12.5 mm layers with staggered joints) on the party wall, again, over the full gable triangle if the property is timber frame. However, whilst it is normal to find lightweight timber walls used in party wall construction, it is also possible that brick/block walls are erected. Another indicator which usually requires a strong torch or lead lamp, is the roof wall plate which exists in the most inaccessible part of the roof.

Wall plates are usually of planed timber (but not always) and can be seen from inside the roof void running along the eaves. A clear view is often obscured by insulation. If this has to be pulled back it should be replaced as it was, ensuring that it reaches the wall plate and leaves sufficient air space above to ventilate the roof.

Wall plates in traditional housing are more commonly sawn softwood. But both types can be used in both forms of construction.

Most loft insulation is 'mineral wool' based on glass fibres or other minerals such as rock or slag. Goggles, a face mask and plastic gloves are recommended when working in confined spaces with this material.

Some systems construct the whole roof at ground level.

Cavity inspection

Gaining access to the cavity may mean drilling a 15 to 18 mm hole which is best made in a mortar joint for ease repair. By inserting an endoscope or boroscope through the hole you can see some details of the cavity. In particular, if the view shows a white or black breather paper or possibly plywood, then the building is timber and brick.

A number of buildings used fibreboard in place of the plywood and breather paper. If inspection shows a dark bitumen impregnated board then, again, the building is probably timber and brick.

Direct viewing with the naked eye is possible, but a larger hole is needed so that daylight or an additional light source can be inserted. Usually it is easier to remove a complete brick, which can be replaced afterwards and repointed, leaving little sign of damage.

STEEL FRAMED STRUCTURE WITH LIGHTWEIGHT CLADDING

The process of investigation follows that of high rise large panel structures, but here the emphasis is on the external cladding behaviour. The inspection is based on figure 3.3

1 Some form of stop should be inserted between floors. This may be:

- missing
- warped
- slipped/loose.

2 The top joint between the window and the floor should be checked for signs of movement.

3 The sill could harbour water which can run into the spandrel panel or down the cavity. Rain can be forced under the sill overhang to penetrate below the window.

4 It is likely that condensation is occurring in the cavity. It may be on the internal face of the inner lining/wall or on the back of the cladding panel. Furthermore, rain in the cavity could travel along the fixings to the inner leaf. If timber battens have been used then rot is possible. Ties that have no adequate protection against corrosion may be affected.

5 All vertical and horizontal joints, whether gasket, sealant or cover strip, should be checked for integrity.

6 A number of points need to be checked with the ties and fixing rails, brackets or clips:

- corrosion – due to water in cavity
- distortion – may be evidence of differential movement between structure and cladding. (One case is of a whole façade slipping down due to inadequate ties.)
- missing ties, clips, etc
- untightened nuts/bolts
- full take up of any movement tolerances.

7 Deterioration of materials in spandrel, especially if spandrel panels were used.

Inspection will be necessary from the inside as well

DETAILS	TIMBER BRICK CONSTRUCTION			
	Yes	Probably	Maybe	No
EXTERNAL Windows fixed to inner leaf				
Compressible layer under window cills				
INTERNAL Plaster work sounds hollow				
Timber lintels over window openings				
Floor joists at first floor level sit on: • a planed timber plate • blockwork				
Gable end in roof void is: • blockwork or brickwork • fully clad in plywood or fibreboard				
Party wall in roof void: • has 25 mm (or more) of plasterboard on it • is made of blockwork				
Wall plate along eaves, in roof void, is planed timber				
CAVITY INSPECTION shows				
Black or white breather paper and/or plywood or fibreboard				

as externally. Indeed the most worthwhile knowledge will be gained from inside. This will entail a degree of taking down to give a good visual and touch inspection, but an indication of the condition of the cavity can be gained with a boroscope.

Access and equipment

The table shown in figure 3.11 gives the common methods of access to low and high rise structures.

3.10 Checklist for inspection of a timber framed house

Visual inspection

There is no substitute for a detailed examination with the naked eye, breaking out the construction where necessary. Where access prevents direct examination, a boroscope can be inserted.

Small mirrors fixed to a probe can also help in

DIAGNOSIS

EQUIPMENT	ADVANTAGES	DISADVANTAGES
Ladders	Cheap and portable	Restricted height to three storeys. Restricted area of working. Safety
Lightweight access towers and mobile scaffolding	Relatively cheap and portable	Requires level, hard surface to be truly portable. Restricted area of working. Height restricted to approximately 4/5 storeys
Mobile hydraulic platforms (including 'cherry pickers')	Very quick and manoeuvrable. Give excellent access to difficult areas. Movement can be controlled from platform	Requires hard paved surface of roadway width. Relatively expensive. Normal type can be used up to approximately 6 storeys
Cradles	No height restriction. Fairly inexpensive. Can be operated by inspector. Electric cradles fairly easy to operate	Requires special rigging if building not already equipped and will therefore be relatively slow to erect. Manual cradles very slow to operate
Abseil techniques with full safety harness	No height restriction. Very quick. Gives excellent access to difficult areas. Relatively inexpensive. Several inspection companies provide this service with photographic and written report	Requires specialist firm. Difficult to carry out forms of testing requiring heavy or awkward equipment
Bosun's chair	Not recommended	
Full scaffolding	No reasonable height restriction. Best possible method of access for inspection and testing. May also be used for repair works	Most time consuming and expensive method. Will restrict access to users of the building. Possible problems with vandalism and burglary

confined spaces. Good lighting is essential for all visual inspections.

Testing

Testing can be divided into two categories: *on site* and *laboratory*.

Almost all site tests can also be carried out in a laboratory – on suitable samples. If possible, carry them out in the field as the results are then available immediately. However, greater accuracy may be achieved in the laboratory.

The majority of on-site testing equipment has been developed for concrete because it is the material used in the majority of non-traditional buildings.

Figure 3.12 gives brief details of the most common types of test.

3.11 Access equipment for high rise buildings

3.12 Common tests used on modern buildings

Other tests

There are other less common tests, for example ultra-sound, radar, radiography and thermography. Details of these are given in the Building Research Establishment Report, 'The Structural Adequacy and Durability of Large Panel System Dwellings': Part 2.

All these tests can be used to identify voids in concrete. Ultra-sound will also give an indication of concrete strength, and radar and radiography are particular useful for detecting the position of metal ties.

STEEL FRAMED STRUCTURE WITH LIGHTWEIGHT CLADDING

TYPE OF TEST	ADVANTAGES	DISADVANTAGES
Concrete strength		
1 Rebound hammer	Rapid on-site method of assessing comparative concrete strengths. The 'mode' (most frequently occurring) not the 'mean' average of several readings appears to give most accurate correlation Good method for site assessment of areas of concrete for subsequent core testing	Results must be treated with caution. Surface preparation and angle of hammer, critical to accuracy. Conversion graphs applied for hammer readings to concrete strengths should be considered only as a very approximate guide For greater accuracy individual hammers should be calibrated by subsequent core testing of samples
2 Internal fracture testing	Generally considered more accurate than rebound hammer, but not as accurate as core testing Newer 'limpit' device is less disruptive to concrete surfaces	Disruptive to concrete surface, requiring making-good on completion Test takes much longer than rebound hammer and is little used by comparison. Less comparative data
3 Core testing	Most accurate method of assessing strength of concrete. Can be also used on masonry and steel/cast iron	Time consuming and disruptive to concrete surface, requiring making-good on completion. Strictly a laboratory test
Concrete cover		
Magnetic cover meters	Most types fairly accurate. Simplest and least expensive are generally quickest to operate and accuracy is acceptable The more expensive models can give indication of bar size	Only work on ferrous metal reinforcement. Readings to estimate size of reinforcement on more expensive models takes considerable time and not very accurate
Depth of carbonation in concrete – chemical indicators	Quick and inexpensive method. Phenolphthalein is most common indicator – shows bright purple colour on uncarbonated concrete. Best used on chip samples but can be used on dust from drilling	Care required on drill samples. Chip samples must be measured at right angles to the surface
Permeability – internal surface absorption test	Best carried out in laboratory	Present equipment is over-complex and not yet fully developed for site use
Chloride determination in concrete	Most accurate by laboratory analysis. Site tests available using test strips or silver nitrate solution	Site test results variable. Particular care required, where limestone aggregate present
Determination of position of fixing, etc – metal detector	Simple and inexpensive. Can detect non-ferrous metals	Can be confused by metal conduits, fine mesh reinforcement and metallic aggregate (ironstone)
Corrosion determination in Reinforcement – half cell potentiometer	Useful site test to determine corroded areas of reinforcement. Silver chloride probes give more consistent results on site than copper sulphate	Results require collaboration preferably by exposure of affected areas but also possibly by resistivity meter
Resistivity meter	Useful check on half-cell readings	More difficult and time consuming to use on site
Stress/Strain		
Measurement strain gauges	Linear variable displacement transducers (LVDT) and vibrating wire gauges used on site Able to measure very small strains. Over types of instrument available	Complex equipment for use on site
Chemical analysis	Useful laboratory tests for cement content in concrete, sulphates, water absorption Analysis of steel/cast iron/non-ferrous metals	Laboratory only

Further investigations

In order to obtain a full picture of the building, its function and behaviour, the internal services will need to be appraised.

Services investigation

It is essential that specialists are brought in to carry out this inspection. For example, lift engineers for the lift, and electricians for the wiring and electrical installation.

The purpose of the investigation is to establish the following points:

- general condition of service
- ease of accessibility for inspection/maintenance/renewal
- compliance with present day standards and regulations
- vulnerability to general conditions of the building's internal environment
- usage rates of the service
- requirements for maintenance
- life expectancy from date of inspection.

Prior to this inspection all available knowledge on the installation and performance of the service should have been obtained from the building's owners/managers and any firms having maintenance contracts.

A checklist procedure for carrying out service inspections on site is given below:

Checklist for service inspections

- position for access
- location and runs of wiring, ducts, pipes, etc
- condition of working parts to equipment
- any additional or altered installations, ducts, etc
- proximity of service pipes, ducts wiring to each other
- ascertain points of penetration through floors/walls
- check for fire safety
- check fire stops/barriers in ducts, etc
- presence and type of heating appliances
- condition and efficiency of heating appliances
- condition of active fire fighting equipment
- state of vertical drain pipes – general condition – joints.

Gas

Following the gas explosion at Ronan Point, London, in 1968, LPS buildings were deemed to be particularly vulnerable to progressive collapse.

To minimise the possibility of explosion, the gas supply to many blocks has been withdrawn. Some blocks still have mains gas and its continued use should be carefully considered.

Even on blocks which were strengthened to meet the new standards of resistance to explosion, evidence has shown that this may not have occurred.

The structure and fabric inspection should identify any problems.

The use of liquified petroleum gas is also banned by many owners, but this ban is difficult to control as tenants can bring in appliances without the knowledge of owners.

Where gas supplies have been terminated, make sure to verify this. In general, view with extra care the inspection of gas pipes and appliances in LPS buildings.

Fixtures and fittings

To build up a complete picture of the condition of the building, you need to report on the state of fixtures and fittings.

Some of the factors you should consider are the type and condition of:

Individual heating appliances	
Sanitary fittings	extent of maintenance/repairs materials
Kitchen units	extent of provision
	style
	condition
	method of fixings
	state of work tops
	conditions of sink top
Floor coverings	type
	condition
	level
Wall finishes	general finish
	any evidence of dampness

This information can be gained by 'random sampling' a number of dwellings/rooms/floors in the building, say around 10% of the total number.

Summary

To check the condition of the internal structure, fabric and services it will be necessary to take apart or open up at selected points. This necessarily means that a void dwelling should be used.

Remember that the inspection of service pipes and wires will cause a great deal of damage within a dwelling.

Common areas should have services running in ducts or shafts and present less of an access problem.

The main problem in inspecting the external structure is access. A number of methods have been described; abseiling techniques are recommended so long as they are carried out by trained and experienced people.

RECORDING INSPECTIONS

The recommendation of the BRE in their report 'Structural Adequacy' Part 2 on frequency of inspections is as follows:

After initial appraisal
- an inspection should be carried out again in one year
- then in two years
- then in five years.

After these three investigations, follow up at minimum intervals of five years.

The initial frequency of the inspections will allow a complete picture to be built up. The building's characteristics can be ascertained and monitored to produce a log book which should not only present the current situation but also show any difference from the previous inspections. The BRE report states:

> 'A technical log should be established and maintained for all LPS buildings required to exceed 25 years service life, from the date of construction. This should contain detailed records of design details, history, assessments and modifications.'

This recommendation should be applied to buildings of five or more storeys (including any basements) with respect to both normal and accidental loads.

For buildings of less than five storeys, BRE recommend that the appraisal for safety should be made for normal loads only, such as dead imposed and wind loads, together with thermal and ground movements.

When inspections are carried out to the structure and fabric, also make an assessment of the services and fixtures and fittings. This can be carried out at five-year intervals and the information will provide guidance to the owner of the maintenance/repairs expectations and costs.

When assessing the life of the building this information can then be taken into account.

If a building's services and fittings are not efficient they can cause a drain on resources and produce unnecessary safety risks for the occupiers of the building.

THE LOG BOOK

The building's log book should contain the information given in the checklist.

Checklist for log book

- address
- location, with a map
- number of dwellings, size and accommodation
- type of LPS
- number of storeys
- any balconies and parapets
- panel type – description
- joint type – description
- connection type – description
- any design variations
- any subsequent modifications, such as strengthening bolts
- carbonation tests – position, results and analysis
- chloride tests – position, results and analysis
- reinforcement tests – position and cover
- ties – cover and bonding position
- accidental load assessment
- normal load assessment.

The log book should outline the details of all inspections at a glance.

One method would be to use a spreadsheet format based on a computer. Figure 3.13 shows a log book page, and is an example of how components could be identified and their condition recorded.

The log book should contain copies of all relevant drawings, photographs and records of maintenance and repairs previously carried out. A4 size drawings should be used. These can be extracted from the log

Component/Item	Location grid ref	Insp 1 date measurements test results	Insp 2 date	Insp 3 date	Insp 4 date	Insp 5 date
External panel with window	A6	Misaligned crack from window .... measure fitting .... joint sealant				
		Differences	Differences No change	Differences	Differences	
External panel on gable end	X4	Carbonation .... Chloride cons .... Reinforcement cover .... Ties	....			
		Differences	Differences Increase in carbonation depth of 2 mm			

book and taken on subsequent inspections. The drawing should show clearly the position and extent of any defects. This can be checked directly on location to see if there has been any further deterioration.

A drawing should be produced which can show clearly where components and items inspected are located. A grid system which is described in the handbook for the external façade inspection would be a useful tool.

Internal components/items/services could be located by dwelling number, floor and position.

3.13 Example of a typical computer based inspection record

COMPUTERS

Although the records of inspections can be stored on computers, there will be hard copies of drawings and other documents.

The computer can be programmed to compare one set of results to another and this can give instant feed back to the owners. Any major problems can be quickly identified. Figure 3.15 shows such a printout produced by the City of Birmingham Housing Department.

KEY ABBREVIATIONS AND GRADINGS

KEY ABBREVIATIONS AND GRADINGS USED ON COMPUTER PRINT-OUT

Block name and record no.
This column indicates the name of the multi-storey block to which the data refers.

CAN (UK) ref.
This column indicates the file number of the initial CAN (UK) Ltd external inspection survey report.

Type of block and ref. no.
This column indicates the form of construction and the original contract number when the block was built. This information is retrievable in the manual archive file.

3.14 Visual and tactile testing by instruments on high rise structures

3.15 Computer analysis print-out. *Housing, Birmingham*

```
              NEW FLATS LISTING BY WORST ELEMENT AND BLOCK TYPE
              PRIORITY IS BY CATEGORY AND NOT BY ORDER IN CATEGORY GROUP
                         Q R PREFIX INDICATES POTENTIAL FOR
                  ISOLATED PERMANENT REPAIR OUTSIDE MAIN PROGRAM REPAIR

A                  B   C   D          E  F G H I J    K  L     M N O P Q R S T U V

ASHFORD               16 111 AFC-FRAM   3  Y Y 0 2F 6.8  47 0.48  2 2 4 - -     - 2

STANDLEYS TOWER 355   D1    AFC-FRAM   1  Y Y 0 2F 6.6  41 0.48  2 2 4 - -     - 2

GOWER                170 257 AFC-FRAM   1  Y Y 0 2F 5.3  50 0.43  2 2 4 - -     - 2

AUDLEIGH              18 118 AFC-INSITU 4  Y Y 0 2F 6.0  62 0.53  2 2 4 - -     - 2

BRAMSBER              53 354 AFC-WATES  1      0 2F ---  41 ----  2 2 4 - - 2 4 - -

BRAMSFORD             55  52 AFC-WIMPEY 3  Y Y M 2F 9.8  46 0.48  1 2 4 - -     - 2

WILLOW               405 349 AFC7       1      0 2F ---  33 ----  - 2 4 - - 2 4 - 3

LEBANON              236 221 FPC55     11  Y   0 2F ---   0 ----  4 3 2 - - 2 4 - -

KEY:
    A: Block Name and Record No.   I: Overall worst element rating   P: Public Balcony Concrete
    B: Can UK Ref                  J: Brickwork mortar mix (1:-)     Q: Public Balcony Handrail
  C/D: Type of Block               K: Overhang (mm)                  R: Private Balcony Concrete
    E: No. in Contract Grouping    L: SO₃% in mortar                 S: Private Balcony Handrail
    F: Test Cont Let               M: Strength                       T: Render Mosaic, etc.
    G: Lab RPI RCD                 N: Condition                      U: Pre-Cast Panels
    H: MDT Risk Code               O: Main Frame                     W: Roof perimeter structure
```

DIAGNOSIS

Contract grouping
Indicates sampling and analysis contract let where 'Y' appears in column.

Test cont. let
Indicates sampling and analysis contract let where 'Y' appears in column.

Lab RPT RCD
Indicates that a laboratory report is available relating to the sampling and analysis contract where 'Y' appears in column.

Multi Disciplinary Team (MDT) risk code
O – Ordinary
M – Medium
H – High
Indicates level or risk to occupier of building from falling masonry, render or concrete.

Overall worst element rating
Indicates the highest risk code for any element or part of the structure as currently known. This column will alter following further investigation and the availability of more information relating to the structure.

Brickwork

Mortar Mix 1 to 5
Indicates the cement/sand ratio identified for the brickwork following the sampling and analysis carried out in the laboratory.

Overhang
Indicates the overhang of the brickwork in millimetres where supported on a concrete nib, obviously this may not be applicable for all blocks.

SO_3 in mortar
Indicates the levels of sulphates contained in the mortar samples taken. Generally 5% and will mean that expansion is taking place in the brickwork.
  Sulphate contents are normally classified as follows:

Low: Below 3% by weight of cement, expected expansion minimal.
Medium: 3–5% by weight of cement, possibility of expansion occurring.
High: 5% + by weight of cement, expansion very likely to occur.

Strength
Indicates potential of brickwork to withstand applied loads, ie wind.

Condition
Indicates generally apparent condition of brickwork as visually observed.

Main frame
Indicates general condition of main frame structure as observed and where tested following sample and analysis survey.

Public balcony concrete
Indicates general condition of public balcony access, concrete floor slabs and balustrade supports where applicable.

H/Rail
Indicates general condition of balustrading and components as observed.

Private balcony concrete
As for public balcony concrete.

H/Rail
As for public balcony concrete.

Render mosaic, etc
Indicates general condition of cladding materials used on the block where applicable.

Pre-cast panels
Indicates condition and fixity of pre-cast concrete panels on the block where applicable.

Roof perimeter structure
Indicates condition of parapet walls or other perimeter structure where applicable.

General condition indicators
1 Urgent work necessary, defects highlighted, could be dangerous, annual inspections being undertaken.
2 Extensive defects evident, requires repair work as soon as possible.
3 Minor defects only, no immediate attention required.
4 Satisfactory.

Summary

Always carry out inspections frequently. It may be necessary to carry these out every year on blocks which are showing signs of defects or distress. This is over and above BRE recommendations.

Try and keep full and comprehensive log books. Make a note of the following:

- all visits must follow a planned investigation procedure and the result must be logged
- an assessment of the components and elements condition
- services and fixtures in addition to structural aspects
- any deterioration noted to be immediately reported back to the building's owner.

USER EVALUATION

It is the engineer's/surveyor's responsibility to ascertain the condition of the building. But in order to understand fully why the defect has occurred or become much worse, other factors need to be considered. These are usually connected with patterns of use and user behaviour. After all it is people who use buildings, and they create the internal environment. Some examples have already been cited, such as excess condensation generated in one area, not being properly ventilated, causing a flow of water into another area. Inadequate control over heating in offices may cause the occupants to install extra heaters, which in turn alter the balance of the environment.

Lack of maintenance by owner or occupier can lead to defects in themselves, for example non-replacement of mastic jointing around windows. This is just neglect by people (although the real problem may be lack of money). If the building is seen as an asset then any neglect should be seen as a cost, and maintenance considered added value.

The engineer/surveyor can glean information from the building's users, if in occupation. This is particularly relevant in the case of housing. A view of the problems of the property by the occupiers can shed light on the reasons for some defects. They may be able to recall previous work or repairs which can indicate the seat of the defect.

BUILDING IN PLACE

The geographic regional and local position of the building can influence the extent and degree of defect. A building on the north edge of a hill will behave differently from one in a valley on the south side surrounded by taller buildings, even if constructed in exactly the same manner. A discussion on the factors influencing building performance was initiated by HARPER (1978) and has been followed up by CHANDLER (1989). Obviously the local weather conditions will be a large factor in the incidence of defects, as will the initial choice of materials. Materials not suitable for the environment will fail quickly. Therefore in any investigation some reference should be made to the prevailing environmental conditions and the building in place. From which direction does the wind blow? Do adjacent buildings affect the wind flow? Is the area subject to extraordinary temperature fluctuations? Is there excessive atmospheric pollution?

APPRAISAL

The appraisal of the nature and state of repair of a structure can be contained to various stages depending on costs and the building owner's requirements.

However, if a full structural appraisal is required, this must incorporate information obtained from the initial inspection, phase 1 and phase 2 investigations. With this whole picture it will be possible to advise the owner of what is required to repair the building, with a reasonable degree of certainty, on repair requirements and likely costs. See diagram overleaf.

DESK STUDY

Objectives of desk studies

Collection of desk study will considerably reduce the cost of investigation work and enable defects to be identified at a much earlier stage.

It may mean the difference between recommending demolition rather than repair. This is because defects were not identified at a sufficiently early stage, or the cost of large scale survey work, including vacating the premises by occupiers, cannot be justified.

The objective is that the technical log for the

DIAGNOSIS

> 1. Desk Study
> 2. Site Inspection and Testing
> 3. Laboratory Analysis

The data obtained from the above three inputs is derived from:

- Basic maintenance inspection
- Phase 1 investigation
- Phase 2 investigation
- Phase 3 investigation

But to be able to appraise, understand and analyse the buildings' condition, the following knowledge or information is required:

- Knowledge of expected behaviour and the structure in service
- Characteristics of the material used
- Original design intentions.

building should contain as much of the original design and construction information as possible, together with a history of the building since construction.

Checklist – sources of history of the building

- original design information including designed floor loading capacity and geological investigations
- original drawings of the structure including general arrangements, elevations, reinforcement details and details of connections
- any site records of construction, Clerk-of-Works' reports, etc
- details of any modifications, strengthening, additions or alterations to the structure since construction
- the maintenance history of the buildings, including results of previous maintenance inspections
- any previous reports, surveys or investigations.

When this information cannot be obtained from the present owner, there are many other organisations which may be able to help. For example previous owners, the original designers, system manufacturers or builders and the local authorities particularly the architects, engineers, planning and building control departments. In fact, all departments may need to be consulted.

Local authorities

This source is always worth consulting concerning ground conditions in the area. It is also probable, even if records have been lost, that there will be staff employed who can remember the original construction.

Tenants

Long standing tenants, building users or caretakers are also useful sources of information, particularly if they possess old photographs of the construction or can remember particular facts.

Limitations of records

These records only form a starting point for the investigation. It is unlikely that they will be complete. As you know, modifications were often carried out on site, so you will have to verify information from the original records by site investigation of the most important details.

SITE INSPECTIONS AND TESTING

The objectives of the work on site is to:

- verify data obtained from the desk study

- examine the cause of defects or apparent defects reported
- check the structure for any further defects and/or deterioration.

Before a thorough programme of investigation can be formulated, it is necessary to understand the nature of the construction being examined and the design intended for load transfer to both normal and accidental loading.

Only then can you make an appraisal of the capacity of the members and connections to support the required loading.

Stages for inspection and testing

A large volume of work is required for a full investigation of any building. It will be easier to control if you break it down into the following stages:

CHECKLIST – Stages for inspection and testing
1. Inspection of overall movements, evidenced by cracking, deflection and/or misalignment of line, level and plumb.
2. Inspection for local movement in individual members of areas using similar evidence.
3. Checks on quality of original construction.
4. Checks on deterioration of primary structure.
5. Checks on deterioration of secondary structure.
6. Safety considerations.

This is covered in detail in Appendix 4 of *Appraisal of Existing Structures*, published by the Institute of Structural Engineers.

Cracking

Cracking is the most common form of damage resulting from structural movement. It is important to plot the pattern of cracking to scale, preferably on full elevation drawings of the buildings. The pattern of cracking is more significant than any individual crack.

Quality of original construction

The quality of original construction and subsequent deterioration can be checking using visual methods, by breaking open selected elements of the structure and by the use of equipment described earlier.

Safety

Safety considerations were included as the last item of the list of site work because it cannot be fully considered until the site works are complete.

Safety is a most important consideration and should be considered in two parts:

1. Collapse or partial collapse of the building due to defects or deterioration.
2. Possible harm caused by falling debris, even small pieces of loose material.

Collapse

Collapse or partial collapse is usually initiated by abnormal or accidental loading. The partial collapse of the block of flats at Ronan Point occurred after a gas explosion. The behaviour of structures under normal loading may give no indication of the likely response to these exceptional circumstances.

A full appraisal must include examination of the load path through the structure and the mechanisms which provide for overall stability and stability of individual components.

Current Regulations, British Standards and Codes of Practice for design for buildings include sections dealing with accidental damage.

It may be difficult to justify existing structures in the precise terms of individual clauses in these later codes. Nevertheless, these should form the basis for engineering judgement.

Falling debris from buildings is much more common than outright collapse, but the consequences may be equally serious. Debris may range in size from small pieces of concrete, aggregate, brickwork or finishes to complete panel sections.

There is usually some indication of distress immediately prior to failure, but this period can be very short. In the case of spalling concrete or brick slips, cracking will be evident before failure, although this may not be visible from ground level. In the case of inadequately tied precast concrete panels, the presence of joints is likely to mask any movement.

In a full appraisal, the extent of inspection and testing should take account of the possible consequences of such failure and the likelihood of failure without warning.

LABORATORY ANALYSIS

Testing carried out in a laboratory is used to confirm results obtained on site and to obtain additional information.

Types of test

The most useful types of laboratory tests are:

CHECKLIST – Types of Laboratory Test
- chloride content of concrete samples
- sulphate content in soil/brickwork/mortar/concrete sample
- cement content of concrete samples
- depth of carbonation of concrete chip samples
- density
- water absorption
- analysis of steel/cast iron/non-ferrous metals
- compression testing of concrete cores
- examination for stress corrosion
- geological testing for subsoil samples.

Expected behaviour of the structure in service

We expect movement in all buildings and structures. For example, sideways in tall buildings caused by wind loading, deflection of spanning members in normal loading conditions, and moisture thermal expansion and contraction. These are all normal forms of movement and should have been considered at the design stage with provision allowed in the form of joints at suitable positions.

Cracking is likely to result from some form of movement. It is part of the appraisal process to determine whether such cracking is harmful to the structure.

THE REPORT

Reports may vary in length from a few paragraphs to a few hundred pages, but it is the content that is important. Always try to be concise and objective. Use clear, unambiguous language. If technical expressions are necessary, explain them.

The layout of a typical report is given below, together with an indication of the contents of each section.

Contents page:	This indicates the structure of the report.
Introduction	The objectives of the report or problems you were asked to solve. Give full extent and limitations brief.
Background	Any further useful information with which the investigation commenced.
Investigation	The work done, details of desk study, inspection and testing. Outline results of testing.
Appraisal	Interpretation of results of investigation. Discussion of alternatives.
Conclusions	Short and concise. Only one set of conclusions. No 'ifs' or 'buts'.
Recommendations	Could range from no remedial work to demolition. Alternatives may be required for consideration.
Appendices	Photographs. Drawings of construction. Sketches of defects. Full test results. Alternative remedial works details, etc.

Writing the report

A report must always start by setting out the objectives. For example, giving the brief for the investigations. Subsequent readers of the report may not be aware of the question which the report sets out to answer. The introduction should give the full extent of the brief and any limitation such as restricted access, no breaking-out of structure allowed, etc. Lengthy reports may have a summary of the main points and the conclusions reached.

The body of the report

The middle of the report should contain background information and, if available, details of the examination, the methods used, type of testing carried out and results of those tests.

It is rare for any problem to have only one answer, and tests may require interpretation.

Include a section for discussion of options as it helps the reader to understand the problem.

The conclusions

Conclusions and recommendations should be as short and concise as possible. They should be objective and based on factual evidence. Always try to avoid alternatives in your conclusions. If enough time was made available and information given, it should be possible to correctly assess the client's needs and reach only one set of conclusions. Recommendations may give alternative methods and costs for consideration.

Checklist

- Obtain a thorough and written brief, stating clearly the purpose and objectives of the investigation. Stick to the brief and discard irrelevant information except where safety is concerned.
- Re-assess the original brief and discuss with the client in the light of information gained as the investigation proceeds.
- Adopt a phased approach to maintain flexibility and enable expenditure of time and costs to be controlled. Plan the investigation and involve inspection and testing programmes designed to fit the brief but allow for further development.
- Determine the type of construction being investigated and understand the anticipated behaviour of the structure, characteristics of the materials used and original design intentions. Check for modifications to the design, changes of use or environmental conditions.
- Recognise the limitations of the types of testing being carried out, access available, time and cost restraints. Include these in the report, with explanation or discussion where necessary.
- Be objective and only report factual evidence.
- Produce a written report in clear concise language. Every report should have an opening, middle and closing section. The opening must give the brief and the closing section the answer to that brief.

The middle of the report should include details of the investigation carried out, its extent and any limitations.

SUMMARY

The strategies for producing effective options are based on sound and informative investigations and initial appraisal. The building owner must be clear on the objectives of the investigation and be aware of the consequences of knowing such information. For example, if the investigation identifies a hazard and this is ignored and subsequently an accident occurs, it is now no 'accident' as the danger could have been avoided. Unfortunately some building owners do not carry out investigations as they are worried about what they might find. This could lead to expensive repairs from funds allocated for other purposes.

The methods of investigations should be suited to the size of the building and its basic materials and form of construction. Some examples of typical buildings indicate how these investigations can be carried out and the scope of information obtained. The building is not an unused inert structure; its behaviour is affected by use and this should be taken into account when analysing the building in place. The external envelope is the interface between the internal and external environments. It must transfer or mitigate the extremes of either environment and interact with the structure. The investigations should take all these factors into account and be presented in such a way as to give the owner/client a detailed, but clear understanding of the building's characteristics, defects and performance over time. Some initial recommendations may be offered as to remedial measures or strategies for further investigations and appraisals.

CHAPTER FOUR

OPTIONS

After the physical assessment of the building, the next stage is to offer and evaluate the options. As this is not solely determined by its physical state the factors influencing the building's future are:

- use of building (meets user demands)
- comfort (up to present day standards)
- ease of maintenance (availability of spare parts)
- cost of maintenance (fuel, fabric, services)
- alternative social needs (priority for other uses)
- land value (potential for alternate use)
- moneys available (by building owner)
- suitability for repair/upgrading/adaptation (type of structure).

These factors will vary according to location of building and time. An illustration of the complex interweaving of the above factors is the developments in the London docklands. Here a mix of development is taking place. Where old warehouses and offices are in a riverside location, and basically structurally sound, they are converted, refurbished into flats (an alternative use) and offices/workshops. Parking facilities and older housing have been demolished in other areas to allow new build, again a mix of housing and commercial development. In the former it is better to retain the structures and in the latter to demolish and rebuild.

The same basic options are available for modern buildings. The main reason for this is that their physical condition is no better than some buildings constructed in the nineteenth century; that this is an indictment of inadequate design and construction in the middle of the twentieth century in the UK is self

4.1 During inspections of high rise structure it may be necessary to remove loose material. *Housing, Birmingham*

evident. What must now be done is rectify this to maximise the continued safe, economic and comfortable use of these buildings and to provide for future demands. In other words, taking the poor state of these buildings, what is the optimum solution for their future life?

Concentrating on the physical condition as the major factor there are four categories of strategies –
emergency
long-term
medium term
short term.

The issues of social and economic demand and land value will be brought into the discussion later. First it is necessary to assess the technological options open to the building's owner.

This literally rains down on people going into and out of the building. Protective measures need to be taken. It may also be necessary to carry out immediate structural strengthening. This may involve the placement of ties or brackets to resist internal explosive forces due to the use of gas in the building. This was adopted for large panel system buildings in the aftermath of the Ronan Point disaster. A gas explosion caused progressive collapes of one section of the block of flats. Subsequent investigation showed that the ties between wall and floor units were insufficient to resist a relatively small internal explosive force. Many LPS blocks were strengthened by fixing L-shaped brackets at wall/floor junctions to give further structural integrity.

EMERGENCY

One of the major problems already identified is the spalling and falling mosaic on multistorey blocks.

4.2 Rain screen overcladding a gable end to give long term protection and increased thermal insulation. *La Courneuve, Paris*

OPTIONS

LONG TERM

This strategy is based on the expectation that the building is to remain in place for many more years. The length of time for this is to exceed 30 years without any further major works – whether of a structural or improvement nature. The basic structural condition must be such that it can be safely assumed that no deterioration will take place to undermine its structural integrity. Furthermore, it meets present-day standards relating to design codes and material assessment. On this basis it will be possible to consider an upgrading or refurbishment to provide an increased level of comfort, serviceability, efficiency and investment related to rent/sale returns. The social and economic paybacks must coincide in this assessment and be judged over a lengthy period. This is very difficult to predict and will be discussed in detail later.

MEDIUM TERM

Owing to forecasted or predicted changes in demand for social need, the life of the building is not considered to be to the original expectancy, for example an office building on the periphery of a commercial zone in an urban environment. Major redevelopment of the zone with significant demolition and upgrading of accommodation. But presently local rent levels are such that a major reinvestment would be hard to sustain and attract the same rents as the inner zone. Therefore an interim refurbishment is the optimum with the options still left open for a complete reappraisal in 15 to 20 years. The present state of the building is such that no major structural or fabric renovation is necessary until towards the end of the medium term period, then it may be open to reconsider anew the options.

4.3 Overmeshing to retain mosaic and render to external cladding on high rise. *Castle Vale housing, Birmingham*

SHORT TERM

Palliative repairs or cosmetic improvements are undertaken in order to give some benefit to the existing users. This is on the understanding that complete redevelopment is the preferred option. The condition of the building is such that deterioration is extensive and will reach a terminal state in about 10 years. Then the only real option is to demolish. In the meantime some alleviating measures are taken to make the users comfortable with the least cost and inconvenience. It is a decision reached with a full cognisance of the need to carry out major works but, owing to financial constraints, has to be delayed. An example of this is the installation of further anchor bolts and the over meshing of external wall panels to retain them in a safe condition. At the end of 10 years the buildings will be demolished.

ASSESSMENT OF PHYSICAL CONDITION

This discussion of assessment will be based on the systems developed by the City of Birmingham for their housing stock, both high and low rise. The city is a large landlord, having approximately 130,000 dwellings in its ownership. Obviously, ascertaining the condition of the property on such a large scale brings about many problems. They have developed a computer run data base which was described in the previous chapter. It is this to which reference will be made as a tool for assessing the state of the modern buildings.

The city has 423 blocks of flats over five storeys high, built between the late 1950s and the early 1970s. These are constructed in a variety of ways but are predominantly in situ concrete (frame or walls) or large precast concrete panel construction. Some have been built in load bearing brickwork. An example of a computer print-out was shown as figure 3.15. A key to the abbreviations is also given. Some further explanations are necessary before discussing the usefulness of this data.

The blocks are described by their name and number (it will be necessary to go to another data base for their location). Initially the blocks were surveyed by the abseiling method and the company's name is stated which carried this out. This will lead to the report reference.

The form of construction is not always from a proprietary system. The uniqueness of the construction must be emphasised. For example, the AFC-Fram is an alternate frame construction using a basic Fram (company name) design and elements. It is an in situ concrete frame with some precast units in walls and floors. The AFC7 is a unique designed in situ concrete frame (designed by the City Architects Department) with brick cladding.

Where a detailed investigation has been carried out on the properties of the materials this is noted. Again these reports can be referenced. Following on from this separate laboratory reports will be available.

The MDT Risk Code gives a crude indicator of possible danger to the people in and around the building from falling debris. This is based on the initial visual and tactile hammer testing of the block. In Birmingham 127 blocks were in the medium to high category necessitating protection for safe access. The initials MDT (Multi Disciplinary Team) refer to a special group of construction professionals set up to assess and develop methods of dealing with the problems of these modern buildings (both high and low rise). The team comprises architects, engineers, surveyors, scientists and housing managers employed by the city. The output from them will be referred to in chapter 5, Documentation.

It may be that only a single element of the construction is badly affected, but that this must be seen to be setting the standard of condition. Of the building 90% may be satisfactory, but the remaining 10%, because of its potential for failure and/or hazard to people, this dictates the severity of the building's condition. The car analogy can be used to substantiate this. The car can be in perfect engine, mechanical and bodily condition except for corrosion to the brake fluid pipes – the car is potentially lethal and the fault must be remedied.

The next categories on the print-out refer to specific brick elements. Bricks will not have been used on blocks with external precast concrete panels designed under a proprietary system.

Condition ratings are given to the overall assessment and to the frame in particular. As mentioned previously, the condition is given to the brickwork – if present. The structural frame will always be assessed on its condition.

The condition ratings are categorised in four groups.

1. Urgent work necessary, defects highlighted, could be dangerous annual inspections being undertaken (this means full visual and tactile mainly by abseiling technique).
2. Extensive defects evident, requires repair work as soon as possible. (This would be expected within 12 months from assessment.)
3. Minor defects only, no immediate attention required. (A longer term view can be taken, with a further inspection after three to five years).
4. Satisfactory – this does not mean that the building can be forgotten – it will require an inspection after five years.

These condition ratings are attributed to the elements identified on the print-out where applicable, such as concrete balconies, mosaics and precast panels.

This analysis is indicative. It does not provide a full scientific basis for action but can show where priorities should be made. It is very useful in a large stock holding where everything cannot be done at once. It will show where a full structural survey is required and what are the major problems of the building.

In the case of individual buildings, a full investigation will always be necessary before any decisions can be made in unclear cases. There will always be circumstances when the structural condition of the building is not considered at all. The future is based on the economic or utilitarian value of the land. For example, a new road route deemed socially necessary may cut through sound buildings. A building in a city centre preventing an economic redevelopment of a larger area may need to go. A case in point, creating much local debate, is the Rotunda in Birmingham. Structurally it can function for at least another 30 years (with some minor repairs and renovations), but to allow a comprehensive development of the Bull Ring shopping area, it would be demolished. The economic value of the land is greater than the continued function of the building. Therefore demolition is always going to be an option, and an option not based on structurally/functional criteria. Some mention of these criteria will be made later, but here we concentrate on decisions for options on the basis of the continuation of the building in its basic structural form.

CRITERIA FOR OPTIONS

The list of criteria can be simply made, but it is a complex interaction between all these which will determine the eventual remedy.

- structural condition
- fabric standards (thermal, fire, sound)
- condition of components (windows, etc)
- roof condition
- condition of internal services (lifts, heating, etc)
- security
- ease of adaptability to new functions
- external environmental improvements
- requirements of tenants/users
- financial constraints
- required future life expectancy.

The interrelationship of these factors carries through in a linear timescale, but a particular item may need to be reassessed and its influence reappraised if its importance changes. Figure 4.4 demonstrates the three stages in reaching a suitable remedy. This may not be the optimum remedy or solution as real life has a habit of redirecting stated objectives. But we must strive from the optimum so that the best solution can be achieved for the building's users, owners and society.

Stage one concerns the bringing together of all the physical attributes to assess the condition of the building. Perhaps this is the easiest stage as this is based on scientific and technological analysis. This is comparative and must not be undervalued nor seen in simplistic terms. There are still no definite answers to questions such as: What is the life of concrete? What happens if bricks are left to spall? How can we measure and predict the rate of corrosion to steel rebars? At best the analysis will show what *has* happened – it will not necessarily predict what *will* happen.

The building should be considered as a whole – internally and externally. The condition may be such, as in the case of some types of PRC houses, that the structure is in such a poor condition that demolition is the only answer. Orlit and Boswell houses have come into this category.

Decision D1 has to be made – whether it is palliative, short, medium or long term repairs/refurbishment, after stage one.

Depending on the type, use, location and form of tenure, it may be necessary to take into consideration the tenants' views and environment factors. The

CRITERIA FOR OPTIONS

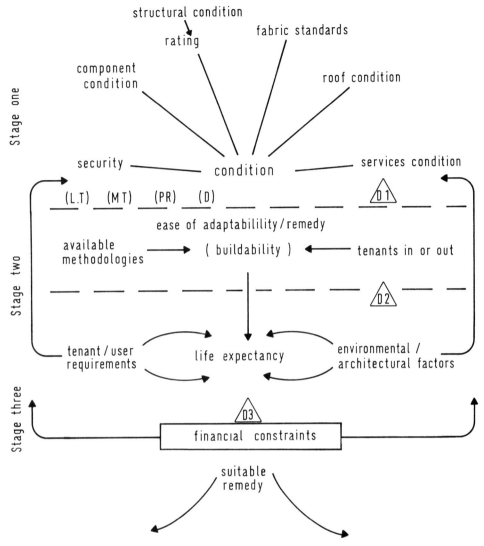

4.4 Interrelationship of option criteria

tenants' views should be ascertained with respect to their observations of how the building functions. This is equally true for housing and for commercial property. In the particular case of a housing estate the architectural environmental factors may be pertinent. The analysis by COLEMAN (1985) is valid here. The condition of the buildings in place and surroundings must be taken into account as part of the final condition evaluation. In some cases this factor alone can influence the decision – especially when combined with tenants' views. Many estates and blocks are deemed 'hard to let' with a large number of voids. Damage to these is common. In other words, tenants (or prospective tenants) have given their verdict on the acceptability of these buildings for comfortable living. They do not like them.

The evaluation at the second stage centres on the feasibility of the measures required to repair or refurbish the building. It must relate directly back to whether or not the efforts are to be directed into long, medium or short term proposals. The physical options need to be considered. For example, whether it is possible to overclad the building; can the services be upgraded; does the roof lend itself to easy reinstatement; can present day functions (such as communications systems) be incorporated, into the structure? In too many cases not enough time and effort has gone into fully assessing the buildability factors. Decisions at this stage should involve manufacturers, trade specialists and others with experience in similar types of schemes. Unfortunately with these modern buildings, even when these conditions have been met, the actuality on site can vary from one building to another adjacent and built at the same time. The case studies described in chapter 8 show the uncertainty of design solutions and carrying them out on site.

Ultimately most technological decisions are constrained by finance. Even where money is available, the concern must be to give value. There are two basic techniques which can be used to provide a cost analysis, namely *life cycle costing* and *value engineering*.

LIFE CYCLE COSTING

A full description of this technique is presented by FLANAGAN and NORMAN (1983). It concentrates on new build but the principles are valid for refurbishment. There are drawbacks to the technique, discussed by BRANDON (1987) and noted by FLANAGAN and NORMAN. These centre on forecasting the real cost of items in use and the need for their maintenance. But the system can provide comparative data which will aid decision making.

The components of life cycle cost, adapted for repair/refurbishment work are:

- total cost commitment to carry out the works, or to carry out particular sections or methods
- short term running costs of the remedial solution – based on the fact that it might be 'plugged on' to the existing building
- the determination of the options which produce the lowest life cycle cost
- the analysis of the present running costs of the building component or service
- the aim to reduce or minimise the running costs after the remedial works.

The main points relating to the application of a life cycle costing analysis are taken from FLANAGAN and NORMAN.

The link between capital costs and running costs must be developed in such a way that the total cost implications of a design decision can be evaluated at the design stage.

The personnel most concerned with running cost aspects of a building are not likely to be involved until the refurbishment has been completed. Estimates of running costs are based on assumptions about future events and these assumptions must be clearly stated.

A life cycle cost approach consists of five distinct components:

- life cycle cost planning
- full year effect costs
- life cycle analysis
- life cycle cost analysis
- life cycle cost management.

Life cycle planning establishes estimated target costs for the running costs of a building or building elements.

Full year costs identify the short term running costs of a proposed building.

Life cycle analysis identifies running costs and performances of currently operational buildings and building components. It is a management tool intended to identify the actual costs in operating a building.

No two buildings or refurbishment methods will have identical running costs, nor will the running costs for any specific building be the same from year to year.

There is a time lag between design and execution (and reoccupation if building is vacated during works) and the availability of reliable data on running costs.

Life cycle cost management is to existing buildings what life cycle cost planning is to the refurbishment building. It is one of the most important areas for the application of life cycle cost techniques. In other words, the value of any remedial or refurbishment proposal must be evaluated with regard to its economic long term viability and cost. It will be foolhardy to embark upon any action without assessing the future cost implications.

A particular life cycle cost equation has been produced in Sweden (GUSTAFSSON, KARLSON and SJOHOLM, 1986). This compares the power supplied to the power losses over time after the refurbishment. It implies that the refurbishment includes an element of improvement to the thermal insulation values.

One side of the equation is the energy into the building, which includes that available from the climate. On the other side is a series of summations involving calculations with factors such as cost of insulation; building cost for climatic shield; cost of heating equipment; running costs on an annual basis; life cycle period; rates of interest and so on. It is a sophisticated model but has easily obtainable data.

LIFE CYCLE COSTING

Use of such a model would significantly improve the basis of decision making with regard to the balance of internal equipment and costs, and external climatic barriers used during refurbishment.

An attempt to quantify financially the options for a Parkinson PRC house is now described. The Parkinson is a precast concrete house which has proved difficult to create an economic and practical reinstatement system.

The Parkinson houses were built in semi-detached pairs. Five main precast concrete columns are on both the front and rear elevations. There are an additional four columns, one in each gable wall and one single storey column central within each house. This is shown in figure 4.6. Ties from front to back of the house are at first floor and roof level in precast beams. Intermediate posts are set at approximately 900 mm (3 ft) centres to frame door and window openings.

Ring beams at first floor and roof levels are formed in twin beam components. The columns have inverted Tee-shaped corbels at first floor and eaves level which support and locate the twin ring beam components. Bolts clamp together the 25 mm (1 in.) wide ring

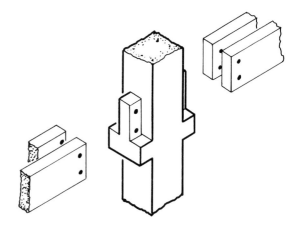

4.5 Typical main column and beam connection. *Parkinson House*

4.6 Layout of columns and beams. *Parkinson House*

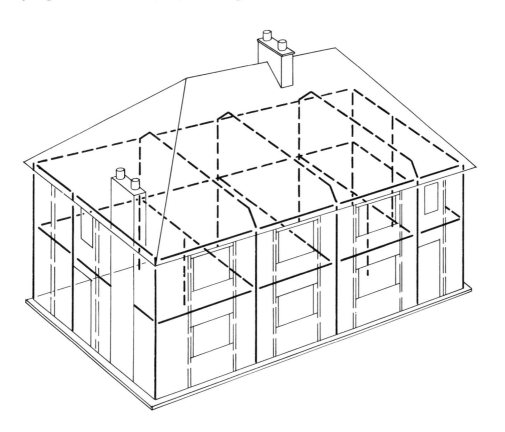

OPTIONS

```
Page No.   1
08/03/88

                       PARKINSONS  HOUSES DATABASE
                              SORTED BY ADDRESS

NUM ROAD       TYP  SR  NR  TR  SC  MC PM EB CL CC FH RS SA RC WC MOD EJ EW KB RW IR COM

 1 BRAMLEY RD   D    66  62 128   0   8  3  3  3  3  3  2  2  1  1   3  2  3  2  Y  3 S Garage at side
 3 BRAMLEY RD   D   100  54 154   5  15  3  3  3  3  3  2  2  1  1   3  2  3  3  Y  3 E
 5 BRAMLEY RD   A   140 108 248   0   0  3  3  3  3  3  2  2  1  1   1  1  3  2  n  1 I
 7 BRAMLEY RD   A   326 110 436   8  32  3  3  2  2  2  2  3  1  1   3  2  3  3  Y  1 I. Diff. chimney settlement
 9 BRAMLEY RD   A   308  86 394   1  32  3  3  2  2  2  2  2  1  1   1  1  2  2  Y  1 I. Diff. chimney settlement
11 BRAMLEY RD   A   287  69 356  13  25  2  2  3  3  3  2  2  1  1   3  1  3  3  Y  1 I. Diff. chimney settlement
13 BRAMLEY RD   D   104  54 158   0  27  3  3  3  3  3  2  2  1  1   3  2  3  3  Y  3 E. Verandah at side,diff chim sett, l/r
15 BRAMLEY RD   D   106  44 150   2  24  3  3  3  3  3  2  2  3  1   1  2  3  2  n  3 E. Garden overgrown
17 BRAMLEY RD   A   232  71 303   3  30  2  2  2  2  2  2  3  3  1   1  1  3  3  Y  3 I. Front porch,dif chim sett
19 BRAMLEY RD   A   302  26 328   0  36  3  3  3  3  3  2  2  3  1   1  2  1  3  Y  3 I.
21 BRAMLEY RD   A   206  56 262   1  31  3  3  2  2  2  2  2  2  1   1  1  3  3  Y  3 I. parquat kit flr,dif chim sett,l/r wl
23 BRAMLEY RD   A     0   0   0   0   0  0  0  0  0  0  0  0  0  0   0  0  0  0     0 I
```

4.7 Condition survey record, computer based.
Parkinson House, Birmingham City Council

beam components to the columns. The reinforced concrete frame is infilled with cavity blockwork. The precast ring beam components stabilise each skin with no use of cavity ties.

Timber joists are used in the first floor and roof.

The computer based data analysis for a Parkinson house is shown in figure 4.7. This system allows a build up of costs to be assigned to each house in order to assess its economic viability for a repair option.

VALUE ENGINEERING

This technique is based on the tripartite factors of *Function-Cost-Worth*. The function of the item/component/technology, should be clearly defined. What does it do? What is it expected to do? The function should be pared to the minimum of two grammatical statements that is a verb and a noun. The verb defines its main function and the noun describes its means. For example, a door can be identified under a number of categories:

verb	noun
protect	room
allow	access
provide	security

Depending on its prime use, the decision of its type, etc, rests. Therefore, 'protect room' might describe a fire door; 'allow access' would denote a simple open/close door; 'provide security' would require a door strengthened to prevent unwanted ingress.

This principle can be applied to more complex elements, for instance a solution for a system of overcladding. It has been diagnosed that the external fabric of a building is not performing adequately. Water penetrates it after heavy rain; thermal loss is excessive; the surface is unattractive in appearance. A number of options present themselves. These are: complete renewal of the existing fabric with similar materials; directly applied overcladding; rainscreen cladding. The three options are analysed using the VE system.

For a detailed explanation of VE refer to DELL' ISOLA (1982) and GILLEARD (1988).

LIFE OF MATERIALS

The science of predicting the life of materials and structures in buildings is only just beginning to realise some answers. Without a precise knowledge of behaviour and its prediction then it is very difficult to evaluate options. Two main approaches are being taken as means of ascertaining behaviour:

1 non-destructive
2 tests/simulation

1 Non-destructive

These assessments are made on materials/components in place without tampering with or altering their composition and/or situation. Examples are; fixing transducers to record movement, stress/strains in an element: monitoring levels of air humidity in cavities; locating steel reinforcement by radar techniques.

The state of development of this type of assessment is shown by this brief description of radar being used to investigate parts of a building (DE VEKEY. BALLARD, ADDERSON, 1989). The Building Research Establishment has been carrying out a series of trials of commercially available radar equipment on simulations of joints in large precast concrete panel structures. The joints were made to simulate site conditions: that is with a variety of situations demonstrating good and bad fitting and connections. The radar operator does not know the condition of the joints. The radar results are then compared with the actual joint conditions. From these the accuracy of the equipment is judged.

On site use has been made of the radar technique with limited success. The beam tends to disperse and cannot clearly locate the different material densities and positions, but it does give some indication of what might be found. Further work is ongoing to improve the reliability of the system. If it can be made to work accurately then a true picture of the actual internal state of joints and components can lead to accurate data being generated for use in assessing the life of materials.

2 Tests/simulation

This approach takes the material or component and tests it. The tests may be to determine:

- exact composition (chemical)
- conformance to standards (to 1S0)

- behaviour in conjunction with other materials/components
- behaviour under adverse conditions.

An example here would be the testing of a cladding system. The glass, mullions, frames, gaskets, etc, will have been tested individually, usually by the manufacturer. These are put together by the cladding company following their design. In order to see if this particular design works, it should be tested. A rig can be set up which holds the cladding and subjects it to wind and rain. This simulates the conditions it will experience in practice on the building. But this is a limited exercise as all conditions cannot be reproduced. For example, an upward wind can be generated due to the proximity of other buildings and features. The simulation rig might not be able to create this – a lack of detail in the design may allow ingress of water from a rising wind. In practice the cladding may fail at the vulnerable points in extreme conditions.

An example of this is presented in the case of a spandrel panel below a window. The original design called for a replacement to the existing panel. The multistorey block was being refurbished, mainly internally, after a change in ownership from the public to the private sector (the previously tenanted flats were offered for sale on the private market). There had always been a problem in the creation of condensation below the windows. The new owners specified a single leaf composite sheet, dry bedded into the existing frame. A mastic sealant on the external edges was used to create a weathertight joint. Within a few months of occupation pools of water appeared below the windows. This penetrated to the flats below and damaged their decorations. The water arose from two sources, internal condensation and external rain penetration. The mastic was inadequate. Also the basic specification of a single leaf created a cold surface upon which vapour condensed. A remedy had to be found for the ineffective spandrel panel which could be fixed externally using the existing framing.

The solution is shown in figure 4.8. An additional insulant was placed onto the existing composite panel. Aluminium sheets with edge strengthening were placed against the insulant and held in place by a foam adhesive/insulant. This foam allowed a variation in panel size to be accommodated as no preformed fixings were required.

This was a case where localised weather conditions acting on inadequate construction components caused a failure and necessitated further remedial work.

It is very difficult to simulate climate, especially where it is created locally (*mesoclimate*) on and around the building (*microclimate*) and at particular positions or details of the building (*cryptoclimate*) (CHANDLER, 1989). During design the possibilities of

4.8 Addition to spandrel window panels

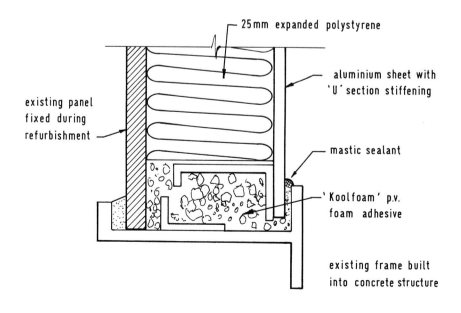

LIFE CYCLE COSTING

4.9 External elevator towers are not adequately protected against the weather. Services and structure to be upgraded. *Chicago, USA*

4.10 Rainscreen cladding system. *Pollockshaws, Glasgow*

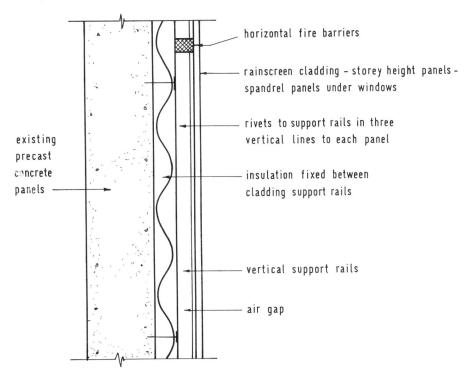

71

OPTIONS

the remedial solutions being just as vulnerable to external problems as the original materials, should never be discussed, or that the materials advocated by a manufacturer do not perform as stated or have not been designed to cover reasonable expectations.

This latter point is illustrated by rainscreen cladding failure.

FAILURE OF REMEDIAL SOLUTION

Multi-storey large panel system blocks of flats were showing the common problems of defective external joints, large heat loss through panels, high condens-

4.11 Replacement of rainscreen panels to gable ends of high rise housing. *Pollockshaws, Glasgow*

ation levels in the flats and flaking/porous concrete. A rainscreen overcladding solution was promoted by a manufacturer of calcium silicate cladding panels. This was designed in conjunction with the client's architect and other manufacturers, namely fixing and insulation. The basic construction is shown in figure 4.11. Within a few years after erection some cracks were noticed at the corners of the storey height panels. The panels were butt jointed to each other and fixed rigidly back to the support rails. Initially it was thought that differential movement between either the concrete panels and supports, or supports and panels was setting up stresses for which there was no tolerance between the panels. Further investigation revealed this was improbable. Meanwhile cracks started to develop along the central line of rivets on a number of panels. The panels themselves came under closer investigation. The panels were approximately 10 mm thick with a fine aggregate surface finish giving an impermeable water barrier. The back and edges of the panels were untreated. Condensation (on the back) and rain (limited on the edge) was being absorbed by the panels. This water caused the panels to expand. There was little or no movement tolerance so the panels cracked. The cladding manufacturer has replaced all the panels on the gable ends of the buildings with sealed impermeable sheets.

Subsequent to this a limited analysis has been carried out on the cryptoclimate behind and within the overcladding. This has revealed that there is condensation occurring in two places:

- on the back surface of the panels (this contributed to the cracking problem)
- between the existing concrete panels and the inner face of the insulation.

Further investigation and analysis is being undertaken. To ensure the individual components of a system are suitable:

- set up simulations testing the design (models)
- require suppliers to guarantee the performance of their products.

The use of a performance approach to design and specification could produce a better solution. This is briefly described here, and again in chapter 5 Documentation.

PERFORMANCE SPECIFICATION PRODUCTION

Setting out a list of criteria to which the ultimate option should meet, allows a range of technical solutions. A performance specification is based on the following stages of development.

PERFORMANCE REQUIREMENT

This is a qualitative statement describing the problem to which a solution is sought. It includes the identification of:

- what is the nature of the problem
- who has the problem
- why the problem exists
- where the problem exists
- when the problem exists.

PERFORMANCE CRITERIA

A set of characteristics must be defined upon which the solution can be judged. For example, thermal resistivity, fire resistance, durability.

EVALUATIVE TECHNIQUES

The judgement needs to be made on a set of common unbiased tests and measurements. For example, noise reduction on a decibel rating. The application of expert judgement may sometimes be the best evaluative technique available.

PERFORMANCE SPECIFICATION

This is the descriptive cumulation of the previous stages in the direct attribution of the problem. It includes the evaluative techniques and range of scores within which solutions must fall if they are to be considered acceptable.

The performance specification can be written in a totally unbiased manner, or it can be written so that it is restrictive, but not as a traditional 'hard' spelt out specification. The 'hard' specification gives drawings, assembly instructions, materials and the like.

This approach is commonly used in the solving of remedial problems, but perhaps not in such a structural manner, nor is it generally in a fully prescribed document. It may be written in general terms only. If this method was adopted more widely it would bring, at least, some sound technical basis for comparison and, at best, produce an optimum solution based on scientific data.

INTELLIGENT BUILDINGS

An option that should be considered in its own right is the degree of technology to be introduced into the remedial solutions. The so-called intelligent building is a concept which takes on board the manner in which the building will be able to function on its own in providing responses to climatic and use patterns. Presently this approach is more common on commercial buildings and is applied primarily to the control of the internal environment. This would only be considered if a major refurbishment was being undertaken, as it would involve a total redesign of the enclosure and internal space and services. The building would have to be gutted in order to reaccommodate the new services and their control systems.

The benefits put forward for a building to take automatic action in response to changes are:

- users do not have to think about controlling the internal environment
- automatic response to users changes in activity patterns
- leads to energy savings in that optimum environments are created in response to number of people and their activity
- the degree of control can be accurately forecast which means that the external envelope can be designed to meet predicted internal behaviour.

Such an approach can produce long term savings, but is only feasible where it is possible to redesign both internal services and external envelope.

FACTORS IN OPTIMAL SOLUTIONS

There are six main areas of concern which will affect the life of a building. It is the interplay between each

4.12 Defective curtain walling and the need to upgrade services means internal and external improvements. *Office building, London*

which will shape the eventual solution. This may not be the optimal one.

1 Physical obsolescence – the period of time between construction and physical collapse. This is rarely obtained as other factors delineate the life of the building.

2 Economic obsolescence – the period from construction to end of occupation when the least cost alternative is applicable. It may be that the land (for redevelopment) becomes more valuable than the rent earned on the building. Or that the lessee/occupier moves to another building for economic reasons.

3 Functional obsolescence – the time for the continuous use of the building for its original function. Modern manufacturing processes are changing rapidly and the building's function may need to be changed, although its structure and fabric remains unaltered.

4 *Technological obsolescence* – occurs when the services and/or components of the building are inferior to present day alternatives. For example, the replacement of air conditioning systems away from water systems using reservoirs.

5 *Social obsolescence* – this is most common on buildings for housing. Due to tenants not appreciating the living conditions; vandalism and local crime rates the housing can become untenable. In some cases this has led to demolition after only 15 years occupation.

There is an argument that the reasons for social obsolescence do not entirely rest at the door of the tenants. Already mentioned is the analysis by COLEMAN (1985) that architectural form and shape can have an influence on tenants' perceptions and behaviour. It may be that housing management policies exacerbate our misbehaviour. For example, placing one parent families in a block can bring problems in its wake. An itinerant male population can abuse the property. Internal conditions of adequate heat may be difficult to achieve owing to low incomes. Social stigmas can be attached to a block of flats, which in turn leads to more abuse. Maintenance may be hard to sustain and the building deteriorates at a fast rate. An instance of this is an 11 storey LPS block tenanted by young people and families with young children. The neoprene joint gaskets were extracted by pulling them down at ground level. This left the joint unprotected with rain water penetrating into the internal wall surfaces.

It was virtually impossible to provide an adequate watertight remedial sealant.

6 *Legal obsolescence* – a local authority owner was taken to court by its tenants on an estate of multi-storey flats. The grounds were that the flats were unfit to live in on environmental health grounds; condensation was excessive, with resultant mould growth on the internal walls; the heating appliances were inadequate (and costly) to provide a reasonable temperature; rain water came through defective windows and faulty joints. The court found on behalf of the tenants that the landlord had to undertake immediate remedial action. This can be viewed as another case of poor management by the landlord.

This leads into a discussion on the effects of management strategies on the life and repairs of buildings, especially housing.

THE EFFECT OF HOUSING MANAGEMENT ON OPTIONS

POWER (1987) has strongly argued that poor housing conditions, vandalism, lack of effective response to maintenance requests, etc, can be vastly improved if management is carried out at local level. This conclusion has been reached after many years working on the Estates Priority Project for the Department of the Environment and the Welsh Office UK.

The condition of buildings, their use and abuse, is affected by the people who live in them. Where improvements have been carried out, repairs undertaken, and then the residents left to carry on as before, the estate soon deteriorates. Two initiatives are required to alleviate this situation: one, the tenants are involved with the major decisions regarding the scope of repairs/improvements required; two, a management office is set up on the estate providing instant access for tenants.

According to POWER, the following lessons can be learnt arising from the installation of a local management office:

1 Local estate offices which were open all day to tenants with full time staff, direct management responsibility and all housing records, brought about an impressive improvement in landlord/tenant relations.

2 Almost all estates were undergoing physical modernisation and adaptation, often incurring major expense. On the whole this reinvestment was successful in rebuilding the popularity of the estates, but it relied heavily on local management to ensure long-term maintenance. Responding to tenants priorities was a pre-requisite of success.

3 Local management of repairs, rent and lettings was only partially handed over to the local office. Where local repairs teams were introduced they were highly popular, efficient and cost-effective. Local lettings brought about a significant reduction in the number of empty dwellings and also

substantial savings in vandal damage, and loss of rent income. However, local lettings and local repairs were introduced on only 8 of the 20 estates (those covered in the surveys).

4 Rent arrears were, on the whole, being contained but continued at a high level, except in four projects where the amount of arrears actually fell.

4.13 Environmental improvements and upkeep in addition to fabric refurbishment. *Doddington Estate, London*

5 Resident caretaking was a vital part of local management and was felt by everyone to be the backbone of the landlord service. Standards rose

in most projects and resident caretaking was recognised as essential in all blocks of flats.

6   The environment of all the estates was upgraded by communal effort, local initiatives involving children and youth, and changes of layout.

7   Most local offices were not given enough responsibility to execute all management functions but they were able to put pressure on central systems and, on the whole, brought about an improved service. The more local autonomy, the more effective was the local office.

8   Beat policing was needed to reduce crime and help curb social abuse. In all projects, major efforts were being made to build up a sense of security and to establish some sense of coherent community control. The local housing office, with the backing of residents and police, appeared to be the key to achieving a normal, peaceful living environment.

9   The residents were directly involved in myriad ways in the development of the 20 projects. This was inevitable when once the office doors had opened. It appeared to be the most effective and long lasting way of ensuring an upturn in the fortunes of unpopular estates.

10 The local management projects had more staff on the ground than previously. But the projects were not as is commonly imagined, expensive. They were affordable by the 19 local authorities, and indeed they brought about substantial cash savings. No local authority questioned the value of the investment on local management.

What should be taken into full consideration during the option stage of decision making is how the building(s) is to be managed and maintained after the remedial/improvement works have been carried out. This applies equally to commercial buildings, although the greatest impact will be on tenanted housing.

SUMMARY

The appraisal of options is a complex activity. It must, obviously, take into account the physical condition of the building in the first place. This is difficult to ascertain accurately without virtual demolition. An early decision should be made as to the main intention, either to go for a long term solution or to consider shorter time span strategies. The physical condition may clearly demonstrate this, but other factors will impinge upon the final solution. It is possible to build up an analytical picture of the building and record this on a computer data base. Some techniques are available to help in assessing the options, such as life cycle costing and value engineering.

The economic, functional, social and legal factors will, in the final analysis, determine the option taken. But in taking that option the management and maintenance of the solution should be built in. In the case of housing in the public sector it is essential to involve the tenants to ensure any remedial works will stand their allotted time span.

CHAPTER FIVE

DOCUMENTATION

It has already been seen that documentation is necessary in the diagnosis and option stages of appraisal processes. The need for documentation does not diminish during the remaining phases. Although it may be seen as the bone of effective thinking and action it does provide an essential record of the condition of the building and the rationale for the chosen options. The full requirements for documentation are listed below:

- initial inspection report
- full structural report – which may include repair/remedial options
- analysis and evaluation of options based on life criteria, social and economic factors
- brief for remedial works
- contact document; bill of quantities; specifications; drawings, etc, for tender stage
- priced specification/bill of quantities/schedule of rates by contractor
- programme plan and intentions
- record of actual work done – measurement, location, cost, by whom, records of tests carried out
- valuation records of payments to contractor
- formulation of building log book.

INITIAL INSPECTION REPORT

The methods of obtaining this initial information is dependent upon:

- type and complexity of building
- in-house skills of inspection.

If the building is tall, or has a complex structure then special means of access and techniques of investigation may be necessary. For example, external access by abseiling.

The skills to carry out this investigation may not be available to the owner in-house. Therefore, an external consultant/company will need to be employed. At this stage it may be wise to go to competitive tender for this work, especially if a number of buildings are involved. The alternative is to negotiate with a consultant with a respected reputation in this area of work. In either case the brief should make clear the following:

- the parts of the building to be inspected
- methods of testing other than visual, for example hammer testing
- type of defects to be recorded
- photography of typical or specific defects
- the notation for reporting defects
- the report format required (possible compatability with computer based data storage)
- the need to remove any potentially dangerous fragments
- the notification of any serious defects immmediately to the supervising officer.

It might be useful to use a video recording with on-site commentary to augment the written report. The object of this exercise is to ascertain where the major problems might be so that the next phase of inspection can concentrate its attention on the important areas. Positions and possible locations should be identified which require further indepth investigation.

CONTRACT

5.1 Photography of defective areas provides essential information. Displaced brick wall. *High rise housing, Walsall*

CONTRACT

A contract will need to be made with the inspection consultant/company. It is better to use a standard form, perhaps with minor amendments, than to prepare one from scratch. Moneys can be retained on the basis of unsuitable presentation/format of the report. This will be important if regulation checks are to be carried out over a period of time. Then the report should be designed so that the subsequent inspections can easily be compared with the previous ones. The objective should be to see instantly if there are any differences.

A 'plan and specification' approach may be most suitable for this initial work. Make sure that basic drawings of the building, together with the specified inspection requirements are issued. The consultants can then price this on a lump sum basis – the contractor preparing his own quantities for estimating purposes. This presupposes that the extent and nature of testing has been decided upon beforehand.

SAFETY

During any inspection the safety of the building's users and public must be paramount. Barriers, notices, access to the roof, person directing people, may all be necessary.

MATERIAL SAMPLING AND ANALYSIS

It is likely that cores, dust samples, etc, will need to be taken to assess the chemical and physical nature of the materials. The documentation for the initial inspection should make clear:

- number of sampling points – making good

DOCUMENTATION

- types of samples
- types of tests to be carried out on the samples
- who carries out the tests
- method of presenting the results
- need for supplementary tests.

The initial investigation report, phase one, should enable decisions to be made regarding the need for further investigations or remedial works.

FULL STRUCTURAL REPORT

Where the phase one report has indicated the need for a comprehensive investigation, then again it must be made clear what are the desired outcomes:

- client's brief – states scope of investigation methods
  – basis for payment
  – mode of report presentation
- expectation of conclusions/recommendations for options.

5.2   Erecting fenced off areas at ground level to keep pedestrains away from works area. *High rise housing, Glasgow*

It is likely that certain parts of the building will need to be opened up. If occupied, it will need to be properly arranged. Agreement will have to be made regarding:

- notification to building's users
- timing of inspection
- who does the 'breaking out'
- extent of breaking out
- method of payment for this work
- need for clients representative to be present.

For example, internal joints between precast large panels may need to be inspected to ascertain the removal of internal carpeting, skirting board and floor/wall finishes. A hole will need to be cut into the concrete. After inspection all damage will need to be

FULL STRUCTURAL REPORT

5.3   Inspection of brickwork and concrete. *High rise flats, Birmingham*

made good. Obviously it is desirable that this is carried out in an unoccupied area, but this is not always possible. For example, the inspection of external cladding fixings. This will not necessarily disrupt the building's users, but temporary safety and weather protection may be required. Sections of the cladding will be removed. A decision will have to be taken as to what is to be replaced.

It is likely that the client will determine the positions of the investigations, based on the initial report.

This report may contain recommendations based on the structural investigation. These recommendations are the indicators which then set in motion the further wide ranging analysis and evaluation of the building. It may seem that this technological analysis provides the optimum way forward, but in practice this forms only part of the solution. This has been discussed in the previous chapter, but it now leads to the means of presenting the preferred options.

OPTION ANALYSIS

The data and information gained from the first and second phase investigations is put into the context of the building owner's expectations. In order to make meaningful comparisons, and to be fully cognisent of the ramifications of any solution, a comprehensive report should be prepared. The information to be included is a follows:

- resumé of findings from structural investigations
- technological alternatives possible based on the condition of the building
- obsolescence factors
- architectural/social factors
- life cycle costs

- energy uses and costs
- maintenance requirements and costs
- methods of financing options.

The relevance and influence of these factors has already been discussed in the previous chapter. Here the emphasis is for a clear, systematic and easily comparable analysis of the possible options.

It is worth reiterating mention of a technique which extends value engineering into a part of the option analysis. It is a further stage in the process of preparing a brief upon which the tender documents can be prepared. KELLY and MALE (1987) have described the value management process. It is a formal approach in the implementation of the value engineering concept. Its advantages have produced considerable savings in the USA and there is every reason to believe that its use in refurbishment evaluation can also bring cost benefits.

BRIEF FOR REMEDIAL WORKS

Once a clear decision has been made, either palliative, short, medium or long term repairs, then a brief needs to be prepared. Two approaches can be made:
1 a traditional specification
2 a performance specification.

In the case of a traditional specification full and unambiguous technical intensions will need eventually to be set down as the basis for tenders, and prior to this, the brief to the designers if further reconsidered under the value management concept. The client's representatives and designers are brought together to discuss the basic intentions. After this the design team continue with their sketch schemes. Again the design team submits these proposals to the Value Management Team. This team, in an intensive workshop situation, then analyse and evaluate the proposals with respect to function, value, performance and cost. This will result in either a revision of the brief or confirmation that it was reaching an optimum solution.

This process can become further complicated in the case of housing estates. The views of the tenants need to be taken into full account. They become part of the design brief and development of the scheme process. When using the value management approach they can be represented in the workshop.

In the case of a performance specification little

5.4 Patched mosaic. Actual areas difficult to ascertain until work starts on site. *High rise housing, Birmingham*

advantage is gained by adopting value management. The responsibility for design and meeting conformance to requirements rests with the individual component/element manufacturers. By going to competitive/alternative companies for this they can be compared with respect to the value management criteria. In other words, they have independently produced alternatives which can be assessed.

The brief for the works must therefore include such factors as:

- clients expectations of design team.
- life expectancy of the refurbished building
- time for design and construction processes
- user requirements
- constraints on methods of refurbishment
- financial constraints.

EXAMPLE

Hollow areas of mosaic facings noted on the defects drawings are likely to be the absolute minimum that will actually be removed when the contractor is required to make such areas safe.

It would be unfair to ask contractors to carry such pricing risks on larger schemes. Here, Bills of Quantities should be provided.

SYSTEMS NEEDING MINIMAL PRE-CONTRACT INFORMATION

For high-rise blocks, a radical alternative can be used.

Contract documentation, by way of a schedule of rates, arranges a competitive basis on which the contractor is selected without detailed knowledge of defects.

Contractor sets up temporary cradle access on site.

Supervising Officer (SO) inspects the structure in detail, the contractor in attendance and the SO instructs the type and extent of repairs to be carried out.

Contractor carries out work

Contractor moves cradle access to next drop or next block.

Documentation and instructions issue by SO allows the work ordered to be valued.

This system is particularly suitable when inspections have been carried out some time before, and further deterioration is likely to have taken place.

A major advantage is that the designers can closely examine the building for themselves to check carefully on safety aspects and also decide upon the palliative repairs most suitable to arrest further deterioration.

Decisions do not have to be made before an inspection on site by the SO together with the contractor.

Insurance cover for the SO will need to be arranged when travelling in access cradles.

Documentation should set out a quantified schedule of rates of the probable means of repair to allow the valuation of physical work when ordered on site.

As the extent of work cannot be accurately forecast at tender stage, it is necessary to allow the contractors to price the means of access and work items separately. Work to blocks of similar height should be grouped together.

This system could be used for Term Contracts (usually let annually) and the contractors are required to provide offers as a percentage adjustment to the issued pre-priced schedule of rates.

This process should not be followed until cost information has been obtained on a number of contracts. However, experience to date shows that different contractors take very different attitudes to the comparison between prices for access and work.

The pricing of access should include the cost of operatives, on the basis that no palliative repairs are carried out during the operation of cradle travelling from roof to ground level.

Repair works instructed should then be priced to include the cost of material and the cost of additional time only of operatives carrying out the repairs.

SOURCES OF INFORMATION

The structural engineer may have sole responsibility for ordering and supervising palliative repairs. In this case the views of the architect need to be sought about the methods of dealing with repairs to any non-structural elements which may cause danger, for example mosaic facing tiles.

CONTENT OF DOCUMENT

The plan and specification methods does not need further explanation, as the type of repair is no different from ordinary small works contracts.

In the case of the alternative schedule of rates system, where the full extent of repairs is not known until ordered, the basis of payment needs to be carefully defined in order to allow the contractor to be reimbursed in line with his competitive offer.

Also the levy of liquidated damages will not be reasonable in these circumstances, and the contract conditions need to be drafted so that the employer does not pay for delays caused by the contractor.

SUGGESTED DRAFT CLAUSES

The following suggested clauses provide a means to comply with the above criteria:

- the contractor will be instructed by the SO of the repairs to be carried out, marked and measured at time of instruction
- the contractor must arrange for his repesentative to travel in the access cradle with the SO in order to receive instructions. The contractor may be required to carry out hammer testing or removal of loose or spalling material under the direction of the SO
- investigation of balustrading or other items may need to be tackled with small hand tools at the time of instruction
- for the procedure of receiving instructions from the SO the contractor is to provide two sets of the following:
  - Measuring rod and tape, marking instruments, 0.5 kg hammer and suitable hand tools. (Two sets are required so that no delay in instructing occurs if an item is dropped from the cradle.)
  - The contractor is to take notes of work and extent instructed by the SO in duplicate form. The instructions will be checked by the SO upon completion of issuing instructions to that cradle drop, signed at the time to confirm. A duplicate copy should be left with the contractor. The original copy of the instructions will be retained by the SO and confirmed formally within seven days to the contractor and the quantity surveyor.

5.5 Work from cradles may be specified. Concrete repairs and brick reinforcement being carried out.
High rise housing, Birmingham

If the contractor considers that the extent of work to achieve satisfactory short-term repairs exceeds that instructed, the contractor shall request the SO to reinspect the works.

If appropriate, the SO shall issue further instructions and the extent of the further work shall be noted.

BASIS OF CALCULATION OF FINAL PAYMENT

The quantity surveyor will calculate the amount of final valuation based on the following:

Access cradles
Payment will be made at the rates offered in the schedule of rates, for provision of access cradles as follows:

1 When a cradle is brought to a block, payment for delivery and positioning will not be due until the SO accepts the 'Certificate of Handover'. The rate is to include hire for the day upon which handover is achieved.

2 Providing and maintaining cradle in position until instructed to move to next position by SO, will be due in whole working days from time of acceptance of 'Certificate of Handover' to time of instruction to move – exclusive of first working day in position.
  Saturdays and Bank Holidays will count as working days even if no work is carried out. (See standing time of operatives for a full definition.)

3 Where the SO agrees that the length of a cradle needs adjustment to achieve access in the next position, further payment will be due at the offered rate, whatever adjustment in size is necessary.

4 No payment will be made after the working day upon which the contractor has been instructed to remove the cradle from a particular block, other than for the removal from the block.
  The contractor is to allow in the relevant rates for the cost of arranging for a cradle to be removed from one block and transported to another or returned to the hirer/owner. EXCEPT that:

5 Should the contractor not return to work within the time allowed under the 'Order to Recommence',

DOCUMENTATION

no payment for maintenance of each cradle per working day will be due for each part or whole day the contractor is in default.

All other plant required to execute the works, required under the preliminaries section, shall be deemed to be priced within the section.

CLAUSES: STANDING TIME OF OPERATIVES

Payment will be made at the rate offered per half hour period in the schedule of rates:

- for each separate period interruption to repair works
- for each full half hour period of interruption of part of half hour, payable only in the following circumstances:

1 The SO instructs or reinspects work carried out from access cradle in the manner elsewhere described, from time of cessation of repair work to time SO leaves cradle – except first hour of instruction on a particular cradle drop (recovery for which may be arranged within the rate of cradle positioning or moving).

2 When cradle is available for use, or all works have been carried out in the current cradle drop position, contractor awaits instruction for SO up to maximum of eight hours per day, except notice of four working hours must be given to SO for first instruction on a particular cradle drop, no standing time being payable until notice expires. Except notice of two working hours must be given to the SO for subsequent instructions on a particular block, or for removal to next expires.

3 The contractor informs the SO and/or they both agree that although the contractor is fully available to work, weather conditions are too cold or inclement for use of the repair materials, provided no other works can be carried out in the current cradle position.

4 The SO agrees with the contractor that although the contractor is fully available to work, the contractor should not be expected to proceed due to weather conditions or other conditions beyond the control of the contractor, which would cause unacceptable risk to his operatives.

The SO may order 'Non-continuance of Work' for reasons given in points 3 and 4, when an interruption has occurred for a continuous period of two hours, after which payment under this heading will cease.

The contractor will not be expected to recommence work until a further instruction has been issued to recommence, which will take effect four hours from the time of the order.

Verbal instructions from the SO are to be acted upon and should be confirmed in writing subsequently.

The contractor must arrange a contract telephone number available during working hours.

No standing time will be payable after instruction for movement or a cradle to next drop of next block, unless the SO fails to instruct within the due time.

PAYMENT FOR REPAIRS

- Payment will be made at the rates offered in the 'schedule of rates', which shall apply to work at any level or elevation of the particular blocks to which the rates relate.
- The schedule of rates is quantified for the reason of evaluation of tenders. The quantities are fully provisional and may not reasonably reflect the extent of work to be carried out or the interruptions which may occur. The rates will be determined by the SO after the detailed investigations from the cradle and by other contract conditions.
- Where noted, the material content of work will be calculated in accordance with the 'Definition of Prime Cost of Daywork', carried out under a building contract, even though the labour and paint content of a particular work item is to be paid at offered or pro-rata rates. For example, repairs to metal balustrading with short sections.

In all other cases, where valuation is to be by means of daywork, the prime cost of daywork shall be deemed to exclude further payment for access cradles.

SUGGESTED SCHEDULE OF RATES FOR ACCESS CRADLES

5.6 Prepared areas of concrete and reinforcement ready with formwork for patching at floor edge.
High rise, Birmingham

SUGGESTED SCHEDULE OF RATES FOR ACCESS CRADLES (ENUMERATED UNLESS OTHERWISE STATED)

- Deliver to any block and erect access cradle with accessories (as described) 9 m long unless agreed otherwise with SO including hire for day upon which handover (as described) is achieved (per cradle).
- Provide and maintain access cradle facilities until ordered by SO to move to another cradle drop (per working day per cradle).
- Move cradle to next cradle drop required by SO including hire for day upon which handover is achieved (per move).
- Change length of access cradle where agreed necessary by SO (per change).
- Dismantle access cradle with accessories, carefully bring to ground level and remove to neck block or return to hirer/owner (per cradle).
- Make good any damage caused to roof surface and structure of blocks.
- Standing time of operatives when interrupted, payable only for those reasons defined in the preliminaries section, calculated per half hour or part of half hour periods (per half hour period).

COMPETITION

Most palliative repairs are basic building trade activities and many small contractors will be suitable for such work.

For high-rise palliative repairs, as long as a contractor can organise cradle movements efficiently and have a team of operatives willing to work at heights, they should be considered for inclusion on tender lists.

Alternatively, abseiling specialists may be able to provide a more competitive price. To allow the SO to order the works without direct access, the abseilers need to carry out a separate drop with a video recorder incorporating an audio system.

The SO can then view the information in the office and issue orders within an agreed timescale. However, it is not possible for the extent of the work to be as strictly defined as from the cradle. This may affect any public accountability considerations.

The prior approval of the employer should be sought to the proposed arrangements. Tenders will not be comparable if contractors are allowed to use either system for the same contract.

Where buildings are very high (over 20 storeys), have an irregular shape and do not have flat roofs, the advantages of abseiling are likely to outweight any public accountability problems.

FORM OF CONTRACT

The standard forms (appropriate to the value of works) should prove satisfactory. A means of inspecting the work before expiry of the Defects Liability Period will need to be arranged by use of independent abseilers.

In the case of the alternative schedule of rates system, the following items need to be amended:

- damages for non-completion
- variation clauses.

ELEMENTAL REPAIRS

You will need documentation for elemental repairs when emergency safety works have to be carried out or funding is so restricted that only one element can immediately be repaired.

With high-rise blocks, this may be an uneconomical way to proceed because the price of access is a fixed cost and would not increase if more were carried out. This is conditional upon the use of the same type of access being appropriate.

Type of documentation

As an example, let's assume that specialists are carrying out bolting work to precast concrete panels. They suspect these panels of not being satisfactorily restrained. Such information would probably have been advised by the abseilers immediately upon inspection.

Where any limited and elemental repairs need to be arranged, the following would apply:

'Plan and specification' may provide a suitable basis for competition for small works. For the specialised works assumed here, the structure to which fixings are to be made cannot be investigated with any certainty. It would be unreasonable in these circumstances to expect the contractor to carry the additional financial risks of estimating on larger schemes.

Also the design solution may not be suitable for all the envisaged situations, and it is better to have a quantified schedule of rates so that pro-rata adjustment of rates can be made at the post contract stage.

Daywork rates should be used if emergency work to a low number of panels is needed as a matter of urgency, and no specialists will provide a lump sum offer at such short notice.

Sources of information

The SO is likely to be the structural engineer in this example. Discussions will probably take place with fixing specialists to check that the anchors proposed for use can be manufacturered within the required timescale.

In an emergency situation, it may be necessary for the SO to accept a standard fixing so that temporary repairs can be quickly arranged to improve public safety.

Content of document

A quantified schedule of rates needs to make provision for:

Quantified schedule of rates
- Dimensional variations of existing structure
- Abortive fixings
- Testing.

Dimensional variations of existing structure

A type of anchor may seem suitable for all the panel fixing requirements, but the strong possibility of existing dimensional variation will require consideration for:

- cutting standard fixing units to length
- manufacture of fixing units of differing lengths (particularly stainless steel which is impractical to cut to length on site).

Abortive fixings

Drilling through reinforced concrete will always be open to the possibility of striking the rebars. As it is unacceptable for any rebars to be severed by drilling, the contractor should be instructed to cease immediately if this occurs. The contractor needs to be ordered to select the position for drilling in conjunction with use of a cover meter, to ensure rebars are avoided in the cladding panel.

The employer will then not pay for any abortive holes in the panel and the contractor will need to make good abortive holes at his own expense.

It is unreasonable to expect the contractor to be able to avoid rebars by use of a cover meter in the in situ concrete frame behind panels or claddings. Therefore provision needs to be made for the probability of around 5% of drillings being abortive.

Further chargeable items need to be included for making good these abortive holes in the cladding panels and if required to the in situ frame.

The SO of clerk of works will need to determine which abortive holes are to be paid for by the employer.

ELEMENTAL REPAIRS

Testing

To prove the strength of restrain fixings in position, the documents need to make provision for a selected percentage to be tested to a minimum torque, in the presence of an SO or a clerk of works.

COMPETITION

For the specialist work, particularly when emergency work is necessary, the use of manufacturers with a fixing division may provide the best service. When time is less pressing, the structural engineer should consult the major manufacturers to check on availability of suitable stock items or the time required to make specials.

5.7 Evidence of insertion of retention bolts. Some of these may not be effective. Allowance must be made for further fixings. *High rise housing, Scotland*

With this information, it is then possible to invite competent main contractors into competition with fixing divisions of leading manufacturers.

FORMS OF CONTRACT

The standard forms (selected in relation to value of works) should prove satisfactory without amendment.

Inspection upon expiry of the defects liability period for medium and high-rise blocks may need to be

89

arranged. A standard order form would usually be suitable for restricted emergency work, under around £5,000.

SAFETY

In the example discussed, the SO may need to set up a sophisticated safety system to ensure that a panel becoming detached from the structure during operations will not endanger the occupants of the cradle.

Also a falling panel will need to be restrained from falling uncontrollably away from the face of the building. The sequence of provision of restraint fixings may need to be instructed to minimise risk.

The contractor should be required to check for cables and conduits fixed to inner skins with a metal detector. This is to reduce the risk of striking live cables. With operatives working from cradles, double insulated drills need to be used.

SHORT TERM REPAIRS

Type of documentation

The repair package will need to correct the defects noted on the inspection reports, and any matters which could lead to danger during the intended period for continued use. While access is available, other unavoidable maintenance can be carried out as economically as possible. This may mean bringing forward maintenance work scheduled for a later date. It will be cost effective to utilise the access for repair.

Both basic building work and work of a specialist nature are likely to be included in such a scheme. Due to the complexity and uncertainty of working to existing structures, main contractors should be provided with bills of approximate quantities for competitive tendering. Provision will need to be made for additional work, because the extent of repair work is likely to exceed even pessimistic forecasts.

Provisional quantities are to be preferred to provisional sums. It is unlikely that a 'plan and specification' basis will be satisfactory for this type of scheme. A pre-priced schedule of rates basis, with competing contractors offering percentage adjustments, is unlikely to provide as keen a tender as with a bill of quantities. This is due to competitive rates being over all schedule items and not specific for restricted items set out in a bill of quantities.

Sources of information

With the benefit of the external inspection report, both the architect and the structural engineer will develop a balanced scheme within the constraints of the brief and the funding available.

Also the frequency of future external inspection will affect the extent of work to be included in this limited scheme. The assessment of the likely extent of repairs will need to be discussed and the recent inspection report considered.

Content of document

Many of the bill items should be measured in the normal manner, but where innovative repair systems are to be used, the standard method of measurement is not likely to provide guidance.

In these cases, the quantity surveyor should provide the bill items in the easiest form, for pricing consistent with the practicality of remeasurement on site after execution.

Elemental bills

Bills should be produced in an elemental form. With repairs being needed in an ad hoc way, this will enable the contractor to price more keenly.

Advantages will be gained later when carrying out interim valuations and in obtaining cost information via elemental cost analysis for estimates on similar later schemes.

Adjustment of the cost analysis for the final account should provide more accurate information when much of the provisional work has been included in the bill.

Repairs

Making full financial provision for repairs by way of measured items is the most difficult matter to be faced. An example of short-term repairs will serve to highlight this problem.

EXAMPLE
Mosaic facing tiles have been identified as ringing hollow when hammer tested by the abseiling inspec-

tors carrying out the external inspections 12 months before.

The report indicates the hollow area to be $0.5\,m^2$.

The mosaic facing is to be cut back to the point where it is solidly adhered. The quantity surveyor needs to make an assessment of the actual area which will be removed by the operative on site.

Two indefinites need to be addressed:

1   How much further deterioration had taken place since the date of the last inspection?

2   How much more mosaic will fall away when the adjacent hollow area is being detached?

It is suggested that on average, a provisional measure in the order of double the area notified as hollow, ie $1\,m^2$, should be included in the bills.

Further dimensional variations of the existing structure need to be considered.

The assessed area where the mosaic has been removed is to be made good with render, but at what thickness? Experience has shown that the expected norm will be greatly exceeded in some areas.

Particular care will need to be taken to ensure that speculative bill items are included for an assessed proportion of the area, which may require a greater number of coats to render to finish flush with the general surface of the remaining mosaic tiles.

Method of preparation and finishing of concrete

To ensure the full extent of repairs are carried out, testing by hammer, for example, will need to be measured to the full area of the material and this may include testing of surfaces after removal of the surface finish. Similarly, the repaired external area may need a finishing treatment to slow deterioration from continuing carbonation.

5.8   Areas for repairs to be accurately measured during remedial works. *PRC houses, Taunton Dene*

Form of contract

The standard forms of contract (selected in relation to value of works) should prove satisfactory without amendment.

Inspection before expiry of the defects liability period for medium and high-rise blocks may need to be arranged as suggested in chapter 3.

LONG TERM REPAIRS

Documentation for repairs to secure longer term continued safe use, will deal with all elements of the structure and finishings to maintain safety for up to 30 years.

You will need to reinspect approximately every 10 years but no repair work should be needed for around 30 years.

An outline of the possible repairs may be encountered is set out below:

Outline of possible repairs
- Repair and refurbishment of in situ reinforced concrete frame.
- Removal of all precast reinforced concrete elements from system built low-rise dwellings and rebuilding in traditional construction.
- Taking down unsound external brick skins and replacement with new, paying attention to new fixings and expansion joints.
- Removal and/or covering of existing applied external finishings.

The opportunity can be taken to carry out improvements concurrently to upgrade the comfort levels of the building perhaps to include:

- improving insulation to roof and new covering
- improving insulation to walls of dwellings
- fitting of replacement double glazed units
- improving heating installation.

The attitude of tenants to the inevitable disruption caused to them during the progress of structural repair works, will become much more tolerant if they know that some improvements to their living conditions will soon be achieved. A scheme involving structural repair content alone is unlikely to seem successful to tenants.

Type of documentation

This type of scheme is likely to be complex with need for the highest standards of repair and renovation to achieve a long-term successful outcome.

Tenderers need to be provided with bills of quantities supported with detailed drawings and specification. Provisional quantities should be included to reflect the likely extent of the repair work. Other methods of documentation are unlikely to provide the control necessary for a major scheme.

Sources of information

The architect will need to co-ordinate the overall scheme with the structural engineer's repair proposals, the mechanical and electrical engineers' schemes and the landscape architect's environmental improvements.

The assessment of the extent of repairs to be carried out to achieve the required standard, should be undertaken in the light of defect reports and sampling and the analysis results by the quantity surveyor, jointly with:

- the structural engineer for structural elements, particularly where areas are currently concealed (eg in situ concrete frame behind facing brick skin)
- the architect for external finishings (eg mosaic facing tile repairs and rendered surfaces).

A recent inspection report of building defects is necessary to accurately assist in the task, as further deterioration since the time of the last report will then be negligible.

Content of document

Elemental bills: these will be appropriate for long-term schemes. The document will have to make satisfactory provision for repairs to rectify both apparent and potential defects.

CONCRETE REPAIR

Concrete repair, to a 'state of art' high standard for long-term repair, should be specified and measured for the following:

- surface preparation to remove existing decoration and friable material (often by grit blasting)

CONCRETE REPAIR

- hammer testing to identify existing and potential defects
- breaking out defects identified by hammer testing and to a sound alkaline base after testing with a chemical indicator
- blast cleaning reinforcement bars which may be exposed, and coat with suitable primer or replace badly corroded bars with new
- application of a bonding bridge slurry to existing structure, prior to use of repair mortar
- filling with repair mortar including necessary formwork and supports
- refurbishment of structure by coating with highly alkaline cementitious render to minimise further carbonation
- painting to minimise future deterioration.

Alternatively, resin injection repairs may be used instead of mortar. Cutting back for mortar could affect

5.9 New openings and reinforcement to jambs in precast concrete panels allowing rearrangement of internal room layout to satisfy new social needs.
Paris

structural integrity, whereas an applied resin can provide a surface finish.

The selection of a particular concrete repair system is a difficult matter which needs to be made by the structural engineer in conjuction with an independent testing laboratory. At the time of writing, no Agrément Certificate or British Standard Specification has been issued to provide guidance.

However, it is recommended that three comparable repair systems are selected. The tenderers are instructed to price on the basis of repair works being carried out by selected sub-contractors, approved by the material manufacturers. The choice of contractor

CONCRETE REPAIR

is important because the preparation work must be carried out conscientiously to achieve a long-term satisfactory repair. The adoption of Quality Assurance by contractor to meet BS 5750 standards should be encouraged.

In order for the employer to enforce contractual responsibilities if repairs fail, the contractor should be instructed to provide drawn details of the position, depth and extent of concrete repairs carried out. The recorded information would be verified at the time of filling with repair mortar by the clerk of works and the drawings provided before practical completion. These should be used by the quantity surveyor as the basis for payment for repairs.

Although, the surface preparation, testing, refurbishment and paint finish can be measured with a reasonable degree of accuracy, the extent of concrete repair is most difficult to predict.

5.10 Removal of brick skin to timber frame houses for repairs to sheeting and new ties. *Nuneaton, Warwickshire*

5.11 Fixing preformed caged insulation to PRC houses. *Swindon, Wiltshire*

This applies even when sampling and analysis reports are available, as the disposition of sampling points can never provide a full picture of the extent of deterioration.

Certain factors need to be considered when attempting to make suitable inclusion in documentation for high quality concrete repairs.

Documentation for high quality concrete repairs

- Grit blasting can have a penetrating effect and will reduce substandard concrete to unbound material. Balcony slabs have been known to disintegrate under treatment.
- The depth of concrete repair after removal of carbonated concrete particularly when it becomes necessary to cut behind reinforcement, increases costs heavily.
- With the contractor being responsible for the long-term effectiveness of the repairs, it will not be

possible to moderate the extent of repair being carried out by the sub-contractor without risking withdrawal of the warranty.

Dependent on general advice from the structural engineer, the quantity surveyor will need to make a reasonable, financial provision for concrete repair without a firm basis (even more difficult where the concrete is currently concealed).

A spread of provisional items at increasing depths (25 mm increments) with size ranges of repairs as set out will more accurately define the items for pricing this high cost work.

EXAMPLE

Cut back and repair concrete surfaces as described:

not exceeding 25 mm total depth

Over 0.50 m²	m²
Over 2.25 not exceeding 0.50 m²	No
Over 0.10 not exceeding 0.25 m²	No
Over 0.50 not exceeding 0.10 m²	No
Over 0.01 not exceeding 0.05 m²	No
Not exceeding 0.01 m²	No

The variation in quality of insitu concrete is so great that the cost of repairing good concrete may be minimal but very poor concrete may reach the point where repairing is uneconomic.

The uncertainty of financial prediction should be indicated to the employer before tender acceptance.

Low-rise PRC houses

Removal of all precast reinforced concrete elements from low-rise dwellings and substitution with tra-

5.12 Reinstatement to PRC house with inhabitants still in occupation. *Airey houses, Taunton*

5.13 Satellite access for external repairs. *Offices, Chicago, USA*

ditional construction is often necessary to achieve a long-term repair.

Designers of schemes approved by PRC Homes Limited (a subsidiary of the National House Buildings Council – NHBC) are granted licences.

This gives an acceptable standard of reinstatement/ repair to allow mortgageability by building societies.

The documentation should require the contractor to:

- prop the existing structure and remove precast concrete units
- provide temporary dust-proof screens for the well-being of those remaining in occupation
- rebuild using traditional construction in accordance with the details of the approved schemes
- phasing of the works so that the time of building operations to each dwelling is reduced to a minimum.

The need for replacement of unsatisfactory sub-fill, sulphate affected ground bearing concrete slabs and foundations will be identified by prior sampling and analysis.

REPLACEMENT OF UNSTABLE BRICKWORK

The expected contraction of a reinforced concrete frame and conversely the expansion of brickwork, perhaps due to sulphage attack, will cause compression of brick panels.

5.14 Installation of new windows during reinstatement of PRC houses. *Cornish houses, Rochester*

REPLACEMENT OF EXTERNAL BRICK PANELS

The documentation for replacement of external brick panels, particularly important for work to high-rise blocks, needs to consider the following:

- the structural frame may not be stable with all brick cladding removed at one time
- the requirement to protect the inner skin whilst the external skin is removed
- provision of temporary supports to existing windows, of sufficient strength for the safety of occupants, where they are to remain in place
- where windows are to be replaced, the operation should be carried out in a short time, to reduce inconvenience to occupants. Temporary windows should be maintained until completion of replacement and internal finishings. Decorations may need to be made good
- repair to inner skins, possible roughly finished, may need repair as well as any services running within the cavity
- the new brick supports may be in stainless steel, standard units not being capable of adaptation to size on site. Various sized units may need to be measured to cater for likely dimensional variation of the structure
- purpose made damp-proof course may be necessary to fit to the profile of the inner skin.

APPLIED EXTERNAL FINISHINGS

Applied external finishings (eg mosaic facing tile, render) sometimes become detached, causing dangerous situations particularly on high-rise blocks.

Long-term repairs to overcome this may be achieved by use of:

- proprietary reinforced insulated render
- rain screen cladding using a proprietary system.

This system does not attempt to seal rain fully from the building face but allows evaporation of moisture through a ventilated fire-stopped cavity overcladding using a proprietary system.

Documentation relating to these types of operations will need to consider provision of the following:

- the method of preparation of full removal of existing finishes. Subsequent failure of this backing must not cause the new cladding to fail. After over cladding, it will not be possible to monitor the condition of the backing, unless a full inspection is carried out
- the requirements of the manufacturer for fixing to the existing structure
- the difficulties of fitting the prorietary product to the profile of the existing building.

These are expensive remedies and other than for improved insulation and aesthetics, a more economic long-term safety measure may be used.

This can be carried out by patch repairing the existing finish and securely fixing metal mesh to the structure so that subsequently detaching material cannot fall.

DOCUMENTATION GENERALLY

Quality control and assurance testing needs to be carried out by an independent laboratory to ensure the designers obtain works carried out to the required specification.

To emphasise further the need for quality, the documentation may require submission of a declaration by the contractor. This declaration should indicate the contractor's acceptance of his responsibilities to make good any latent defects within a specified period at no cost to the employer. This declaration will clarify matters if it is necessary for the employer to enforce its right of repair or to simplify proceedings at law.

Designers will always attempt to improve methods of repair at lesser cost, but it is hoped that the above comments will assist when analysing the difficulties of provision of documentation for any long-term repair scheme.

FORMS OF CONTRACT

The standard forms of contract (selected in relation to value of works) should prove satisfactory without amendment. Inspection before expiry of the defects liability period and at least again before the period to which the contractors declaration applies, should be arranged.

DOCUMENTATION

SAFETY

The comments already made will apply, but in addition, with particular reference to high-rise blocks requiring full scaffold to enable repairs, some further points should be considered:

- temporary entrance canopies to protect building users from falling material
- nylon mesh protective sheeting, attached to scaffold in the area of building operations where above adjacent footpaths

5.15 Overcladding replaced to 1960s in situ concrete framed offices. *London*

5.16 Sheet protection to those floors being worked on. No boarding to the lower scaffolding. *High rise housing, Birmingham*

- all scaffold boards and other loose items to be tied down due to the possible effect of wind pressure at high levels.

SAFETY

SUMMARY

In planning and producing documentation for inspection and repairs the following need to be considered:

- scope and aim of inspection
- scope and aim of repair
- limit of liability upon the contractors.

It was seen that inspection can be a multi-staged activity. Each stage requires its own brief, contract clauses and documentation.

The inspector must provide a clear and accurate report of his findings, and the inspection must be carried out safely.

When carrying out sample analysis, the contractor should assume that a standard contract will be offered. Testing and analysis should be put out to tender and clear instructions need to be given on frequency and position of samples.

The scale of testing will determine the type of costing basis, such as lump sum or schedule of rates.

Palliative repairs can be carried out under a plan and specification or schedule of rates type of documentation. The use of Term contracts is discussed and suggested draft clauses are presented.

When dealing with elemental repairs the plan and specification method is advised. The standard form of contract, with amendments is most suitable. Special note must be accorded to retention of structural integrity during and after the repair.

Bills of approximate quantities would be helpful in costing/tendering for short-term use repairs. It may be necessary to provide bill items in the easiest form or special repair systems. Standard forms of contract, related to value of work are most appropriate.

Long-term repairs should be based on full bills of quantitive specification and drawings. Reference should be made to quality assurance. In some cases it may be appropriate to use the Institute of Civil Engineers (ICE) forms of contract rather than Joint Contracts Tribunal (JCT) forms. Safety aspects are of paramount importance and should be made prominent in clauses and descriptions of work.

PROGRAMME PLAN AND INTENTIONS

Under the JCT 80 standard form of contract, the contractor does not legally have to provide a programme of works. But where the building is to remain partially or fully occupied then it is in the interest of both client and contractor to develop and work to a programme. This should be more than a passing reference to the bar chart pinned on the site office wall. In the circumstances it is the building's users who will need to be kept informed of progress and intentions, especially where access will be required to their areas.

Sequencing and the integration of specialist trade contractors can best be shown via a programme, preferably a network analysis with a simple bar chart demonstrating the main activities. The sub-contractors should be made fully aware of their times on site by being furnished with a copy of the programme.

There is a strong case for the programme being made a part of the contract documents where building users remain in place.

RECORD OF WORK DONE

All too often the records of the initial construction do not exist. Even if they do they may not reflect the 'as built' construction. When carrying out remedial works, both the client and contractor should keep accurate records. These should be agreed as showing a true picture of what was found, how it was remedied and the final solution. The client requires a record:

- as a statement of what was done
- for future reference when carrying out further anticipated maintenance
- in the hopefully remote, possibility of failure of the remedy.

The contractor requires a record as a statement of what was done:

- to show in detail what was done and paid for
- as feedback for future similar contracts
- to assess what was profitable.

The records of the refurbishment should be kept in their entirety. Test results, problems encountered, who carried out what are generally the province of the contractor. The client should take independent records, but can ask the contractor for copies of test certificates, etc.

The individual concrete repairs will be recorded under the different depths (this one is 25 mm). The

5.17 Overcladding to PRC houses sequenced to give a follow through of work to the different trades.
Woolaway houses, Taunton

increments will have been decided by the specifiers based on the initial inspections. The record should include:

- exact location of repair
- full measurement (area)
- any particular problems (eg bent reinforcement found when exposed)
- photograph of cut-back concrete before repair
- depth of cover to reinforcement
- depth and type of reinforcement
- depth of breaking out
- condition of sub-strata
- type of repair carried out.

VALUATION RECORDS

Most construction activities come to be measured in monetary terms. It is useful to keep records of what was paid, to whom and why. This will provide feedback on the most expensive items and stages of the work; what was undervalued in the estimate and what was missed in the specification and/or bill of quantities.

These records can be part of the total contract records, but they may need to be specifically requested if the financial control of the project was in the hands of a internal or external consultancy.

LOG BOOK

At the end of the day, this is the most important document. This can be made up from a number of separate records covering:

- general description
- history of building (including previous maintenance)

- exact description of structure, load disbursement, etc
- details of services (lifts, waste disposal, etc)
- details and references to investigation reports users/clients comments on patterns of use, problems, etc
- outline of findings from investigations (summary of test results, etc)
- appendix on the extent of remedial works carried out
- detailed appendix showing results of inspections prior to and after remedial works
- performance of building with regard to energy use, maintenance costs, behaviour of elements and so on.

The log book is best stored on a computer. This will allow easy updating. Programs can be made which can give an instant comparison between sets of data. For example, a year by year energy cost comparison; a comparison of a panel inspected over a regular period of time.

It will also be necessary to keep paper documents such as drawings, belts of quantities, etc. These can be microfiched for easy storage and retrieval.

If the refurbishment contract documents (drawings and specifications) have been produced by computers then it will not be necessary to repeat hard copies. What must be done, though, is to make sure the 'as done' records are fully entered into the system. A simple reference on the original documents can alert the viewer to the existence of the 'as done' documents.

SUMMARY

The last document described in this chapter is the most important of all – that is the Log Book. All the other documents, in their various forms, contribute to the formulation of the history of the building.

The initial inspection report will, if available, have referred to the original (and subsequent) contract documents. Therefore, these will form the basis of any newly structured log book.

The full structural report will given information on the form of structure and the manner in which loads are distributed.

Future reference may be needed to evaluate the performance of the selection option. Therefore a document should exist which described the possible solutions and the criteria upon which the one chosen was based. The performance cannot be measured unless the initial expectations have been laid down.

A clear brief should be generated for the design team. This in turn may need to be re-evaluated in the light of a value management exercise.

Whatever term of repairs/refurbishment are adopted, then different contract documents may be required. The scope and nature of these should be based on the constraints of the technologies of the repairs.

After the tender has been accepted then the priced schedules/bill of quantities are checked and retained by the client. These will provide a basis for future comparison.

The programme can be the most important document for all parties, especially if the building is to be occupied during the works.

Records during and after the works, in detail and general, should be made. These will be far more extensive than those generated by new build as they will have to record the exact extent of repairs.

Cost records can be reconciled with the priced schedules or bills of quantities.

Finally, to return to the beginning of this summary, the log book should tell it all, albeit, in some instances, briefly, but with reference to the other documents. This log book should be updated to show any further activity or concerns with the building.

CHAPTER SIX

COST FACTORS

The viability of any remedial or refurbishment work has to be judged, in the final analysis, against demolish and redevelop. This question will be discussed later in the final chapter. At this stage it is assumed that a decision to carry out remedial work has been taken, therefore the cost implications are considered under the following categories:

- comparison of methodologies
- comparison of materials
- pricing strategies by contractors.

COMPARISON OF METHODOLOGIES

Access

On tall mutli-storey buildings the problem of access poses a difficult cost comparison. A full scaffold is the most expensive. This can be reduced by working only on five or six floors at a time, thereby saving the cost of fully boarding all floors. But it may be necessary to enclose the building completely to:

- give protection against the wind (stop boards lifting) and the weather (enables work to be continued in adverse climate conditions)
- provide protection to the public from falling items, cleaning processes (grit blasting) and dismantling procedures.

Scaffolding will need to be tied back to the building. This is not a random positioning but is based on a full structural analysis and design to ensure the scaffold will perform safely. Therefore design costs need to be taken into account.

During the works all bracing, ties, etc, should be maintained. Also it may be necessary to provide access from the inside of the building; this is usually possible if the building is empty. It may also be possible, in this case, to utilise the existing lifts for personnel movement. It is best to provide a separate external hoist so that materials do not have to be carried inside the building.

The advantages of using scaffold is that it provides a permanent, strong and relatively large area for the storage of materials, easy pedestrian access to the whole façade, and allows delicate operations to be carried out in steady conditions.

Vertical platform hoists (satellite towers) will need to have their towers tied back to the structure of the building. The length of platform can be enough to spread safely across an elevation. Usually two separate platforms will be required to provide satisfactory coverage. This system is less expensive to set up and run than scaffolding. It does not provide any weather protection to personnel or repairs. The platforms need to be moved up and down to reach the work position. This will obviously have to take place when the operatives need a meal break or to fetch a forgotten item. Supervisors cannot directly and concurrently check operatives unless they stay on the platform with them. Great confidence has to be placed in the operatives getting it right first time! The final tests must demonstrate that the work was done correctly. When these tests take place the platform cannot be used for production. It has to be taken out of service to allow the inspectors unhindered access.

This same disadvantage is also true for suspended cradles. Access is limited to two people, with perhaps a third. A further disadvantage over a platform is that most cradles require two people to operate them. Each end of the cradle has its own independent motor which requires individual control: a platform can be controlled from one position. Cradles are the

least expensive method of access but are the slowest and most restrictive in scope of operations. They are vulnerable to strong winds which will preclude work being done, and without adequate tying can swing round from one elevation to another.

Despite its comparative high cost, scaffold does offer the greatest versatility in production techniques.

When platforms and cradles are used, large plant (such as grit blasting machines) cannot be carried, and heavy components cannot be carried easily. The sequence of work is governed by the scope of access and the number of times the work place has to move to the activity. On one repair job to brick panel walls

6.1 Temporary tie fixed back to structural concrete with new brickwork built up around it. This will need to be made good after removal of tie. *High rise housing, Stafford*

6.2 Removal of balcony panel at each floor level to allow access onto the scaffold. *High rise housing, Stafford*

COMPARISON OF METHODOLOGIES

11 operations had to be carried out which meant that the cradle had to be lifted up 11 times. (This was not the actual number of times of lifts, as some of the operations took about a day, which meant the cradle going up and down at least ten times for that one activity.) Much time can be wasted by this constant need to travel to the work.

Abseiling techniques are the cheapest of all, but give extremely limited access for repairs. This technique is ideally suited for inspections and tests, not to carry out any real remedial work other than taking off loose materials.

The scale of repairs/refurbishment will normally dictate the most productive form of access, although certain stand alone palliative repairs are best carried out from platforms or hoists.

EXAMPLE

- inserting anchors or ties from an external leaf to the structure
- placing a mesh overcladding to a façade
- carrying out minor isolated concrete repairs.

Where items such as replacement windows need to be put into place a scaffold provides safe access.

In costing access the following factors should be considered:

- hire rates for the items
- fuel/maintenance costs
- time required (need for weather protection over winter months)
- methods of hoisting/moving materials/components
- restrictions on times of working
- time for personnel to reach work position
- number of times for personnel to go to work position

6.3 Satellite platform hoists need to be tied back to the structure. *Office block, Birmingham*

COST FACTORS

- productivity rates/ratios – scaffold
  - platform
  - cradle
- total labour time required for each activity – then total times for each floor.

The balance should be made between:

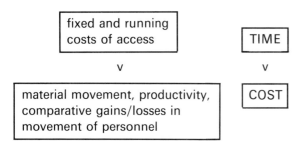

Sequencing

The key to cost effective repairs and maintenance is in the correct planning of the sequence of operations. This is equally true of new build, but there are some major differences.

In most new build the sequence tends to follow a natural pattern, starting from the foundations, superstructure, roof, then internal finishings, etc, either following upwards or coming down. In repair or refurbishment it may be desirable to limit areas of work, especially on large buildings. This may be imposed because of the users remaining partly in occupation. The need to provide some weather protection to exposed elements can restrict areas of work – the cost of protection can be unacceptably high for a complex building. For repairs to structural members, only a few at a time can be worked on so that structural integrity and safety is not compromised. These may have to be worked on in a sequence within the already limited area.

Where low rise buildings are being reinstated, for example PRC houses, then the removal of the elements needs to be in a controlled sequence, with adequate propping and bracing: one wall at a time to be taken down and replaced with the new wall, but the loads not transferred from the props until all the new walls have been built.

The need to go back to an area or house a number of times for the same operation means that careful integration and communication is required between the supervisors and trade specialisms. Any delay in following up work to a building or area will cost money. This factor contributes to the high cost of carrying out reinstatement work to only a few houses. Continuity of work for the specialist trades is difficult to achieve on a small number of houses. It may mean that the trades have to leave and be brought back to the site; this incurs extra cost.

The sequence of work is determined by:

- users remaining in occupation
- structural integrity/safety
- protection of exposed elements
- scope of remedial works – different trades/activities
- means of access (on multistoreys working from the top means scaffold can follow down)
- curing times/striking times for concrete repairs.

COMPARISON OF MATERIALS

With much innovation and adaptation of materials for new uses objective comparisons are difficult. Each product manufacturer will extol the virtues of their materials. The specifier and/or contractor has to assess the suitability of the material, where the ultimate criterion may be cost. In other words, the cheapest is bought. This can be satisfactory provided the product will perform adequately and give the required life cycle.

A full financial analysis of comparative materials should be carried out at the design or pre-tender stage, using life cycle costing techniques. Checks should be made that the manufacturing process will produce the correct standards. Visits to the works should verify the capability of the supplier to provide:

- the right standard
- quantity
- at the required time.

It is likely that the comparative price of alternative materials will vary according to time. Therefore at one time product X may be more expensive than Y, and at another time vice versa.

Another factor that should be taken into the cost calculation is the lead time or availability of the materials. Indefinite lead times can produce high costs due to late start times for the relevant activities.

COSTING AND TENDERING

6.4 Only two people can work from a cradle. Space and movement is restricted. *High rise housing, Birmingham*

COSTING AND TENDERING

Materials

The materials for the repair may or may not be specified by the employer or his consultants. If they are not, a process of selection will need to be undertaken.

Concrete repair materials are manufacturered by a large number of companies, each having an extensive products range. Further technical knowledge about these materials will almost certainly have to be sought.

If the materials are specified, permission may be sought to use an alternative. This can make a tender more attractive to the client representing a saving in contract cost or contract period.

On submission of an alternative, the repair method must be at least equal to the one specified. But beware of any additional contractual responsibilities that you might be taking on if an alternative is adopted.

Selection

To assist in the selection of materials and identify cost implications the following list has been compiled:

Suitability
Check with the manufacturers that the materials are suitable for their intended use.

Availability
Ensure that the materials specified or otherwise are available in time to meet the requirements.

Previous experience
You or someone within your company may have experience of working with the selected or specified materials before; this may well help with the correct selection.

COST FACTORS

6.5 Insulation, mesh reinforcement and cement/sand render as overcladding. *High rise block, Birmingham*

Technical information
Manufacturers will generally supply technical data free of charge.

Many will also arrange to have their own technical staff visit you to provide assistance. They will supply information or covering capacities and average labour outputs, etc.

Compatibility
Always check that if a repair material is to be overcoated, the two materials are compatible. It may be that the repair material will require treating before any other application can be made.

Curing
The curing time of repair materials may well be a costly one. It will also certainly have significant effect on access costs, with scaffold standing while curing takes place. The effect on labour outputs and the contract programme will also need to be considered.

Packaging
The repair materials can usually be purchased in various size packages. Suppliers generally give greater discounts when large amounts are purchased and delivered in one consignment. However, this may not be cost-effective. Waste, storage and shelf life must be considered.

Purchasing a larger number of small packages as opposed to a small number of large packages may well prove to be easier to distribute around the site and impose lighter loads on scaffolding.

Waste
The waste factor can be very expensive. Operatives must be made aware of setting times of repair mortars. The amount of mortar mixed must not exceed the output the labour can sensibly achieve.

The time that the package is opened its workable life must be carefully monitored. Materials purchased in pre-mixed packages can save on waste.

COSTING AND TENDERING

Storage
The manufacturers recommendations must be followed to avoid waster. All material should be kept undercover and most will have to be stored above a stipulated temperature.

Considerations must be given to the time of year.

Shelf life
Many materials only have a short shelf life and should be ordered accordingly and used in strict rotation.

When taking into account discounts for purchasing large quantities, shelf life is often overlooked and becomes a very expensive waste percentage.

Weather restrictions
Many repair mortars cannot be used when the temperature falls below 5°C. The cost of providing temporary weather protection and heating must be compared to lost production, if working during the winter period.

Health and safety
Operatives using these repair materials may need protective clothing, ie gloves, goggles and overalls. Some materials do burn when in contact with the skin. Manufacturers should always be consulted about this matter.

Testing

The testing requirements will generally be set out in the contract conditions, such as the number of tests, where and when.

Test instruments
It may be necessary to purchase specialist test equipment. These instruments can be expensive. The

6.6 New render to floor edge beam. *Multistorey flats, Stafford*

operators generally require some degree of training and experience, so consult manufacturers.

Test cubes

With the very high cost of the repair mortars, the actual cost of the materials used in the test cubes can be significant. The number of cubes required must be costed accurately.

PRICING STRATEGIES BY CONTRACTORS

Examination of tender documents

When the contractor receives tender documents, they are always examined in detail. The conditions of contract, bills of quantities, tender drawings, specification and all other supporting documents will be cross-checked.

All members of the estimating team, including those involved in programming, temporary work and purchasing, will also need to examine these documents.

Conditions of contract

Any unusual features in the conditions of contract need to be identified and the consequences of such condition noted, so that costs where necessary can be included. For example:

- are non-standard forms of contract used?
- what conditions of standard forms are amended?
- are there any non-standard payment or retention provisions?
- what performance bonds are required?
- can the insurance requirements be met by the contractors standard policies?
- are the liquidated damage charges acceptable?
- have amendments to other project information been made? ie items included in the bills of quantity not measured in accordance with the Standard Method of Measurement.

All items listed will have particular bearing on the costing of repairs to non-traditional buildings.

Specification

The specification is to be read in depth. The use of specialist products should be identified and if necessary further technical data obtained from the manufacturers.

Before cost build ups are begun a full appraisal must be made of the materials characteristics. For example, pre-application preparation, mixing proportions curing times, application thicknesses, etc.

The specification may state that certain operations have to be carried out by specialist subcontractors. Enquiries will have to be obtained.

Bills of Quantity

The bills must be carefully analysed and divergences from the SMM noted. In many cases of repairs to high-rise concrete structures, the bills of quantity will be measured provisionally.

An assessment of these quantities to establish their accuracy will be made. An over or under measure will have serious implications on both the cost of the programme period:

- loss of profit (if over measured)
- preliminary costs
- labour resources
- discounts offered by manufacturers dependent on quantities ordered
- hire charges on plant, scaffold, etc.

Tender drawings

A study of the tender drawings will build up a picture of the project. This study must highlight the cost significant items and draw attention to areas of caution or risk.

From the drawings ascertain the following:

- site boundaries
- means of access including approach roads and entrances, any restrictions on access
- location of services
- position of height of buildings to be repaired in relation to other buildings on the site
- position and depth of existing foundations
- temporary structures
- working areas
- restricted areas
- general arrangements for the project
- work necessary to any adjoining structures or property
- any specifically designed temporary works
- any proposed construction methods.

PRICING STRATEGIES BY CONTRACTORS

6.7 Foundations for new external walls may be deeper than expected. *Wates houses, Bristol*

A site visit and/or visit to the consultants will be absolutely worthwhile.

Tender programme and method statement

A meeting early in the tender period between the estimating team, planners and contracts management is essential. Initial thoughts on the construction methods and techniques can be discussed and set out so that all members of the team are following the same approach to the project.

When deciding upon a construction method the following points must be considered:

- site, location and access
- degree of repetition
- shape of the building to be repaired
- availability of working space
- adjacent buildings, structures, etc
- previous experience on work of a similar nature and location

- availability of suitable labour
- availability of materials, specified or otherwise
- extent of specialist work and its relationship to the overall project
- amount of work to be subcontracted
- plant requirements
- degree that design indicates construction method, ie restraints, formwork, striking lines, curing time, special sequence of construction.

Decisions concerning the resources to be used on the project should take into account:

- location and availability of labour and management within the company
- cost of recruiting additional labour, its availability, quality and quantity
- amount of work to be subcontracted
- availability of materials especially those specified
- current and future projects in the area which may affect the supply of basic resources
- quality of workmanship
- specialist requirements
- materials handling on site, storage, distribution and waste, overall time spans and the seasonal affects on the construction method
- quality and complexity of the work.

Alternative methods of construction and sequences of working should be evaluated at this stage and the intended method of construction be proposed.

Preparing a tender programme for repairs to non-traditional buildings is essential. The programme will be used to:

- verify the dates given for construction can be achieved
- maximise the efficiency of resources
- price the time related elements contained in the estimate
- establish the method and sequence of working.

The programme will identify the parameters of the project and the main resources needed. It will provide key dates for major portions of the work and the basic information which can be used when obtaining quotations for materials, plan and work to be subcontracted.

A clear basis now exists on which the construction method and sequence can be developed. Clarify any queries with the consultants and during the site visit. When the queries have been resolved the tender

COST FACTORS

6.8 Bringing up new brick wall after wall ties have been positioned. *Multistorey flats, Birmingham*

programme and method statement can be finalised.

The method statement will outline the sequence and methods of construction upon which the estimate is to be based. It should indicate how it is intended to deal with the elements of work and should highlight areas where new or difficult methods are necessary or intended. It will be supported with details of cost data, gang sizes, plant requirements and supervision requirements.

The purposes of the method statement are:

- to establish the principles on which the estimate is based
- to familiarise construction personnel of the resource limits which have been allowed in the estimate and to describe the method of working envisaged at tender stage.

The final tender programme must clearly identify the construction of the work. It will be used to check and balance the resources needed on the project against those already contained in the estimates.

SITE VISIT

A site visit can be made, and it will almost certainly be necessary to visit more than once; after preliminary assessment of the project has been made and again when the programme and method statement have been produced. Each visit will undoubtedly highlight other problem areas.

The following points should be noted when visiting the site, over and above those for a conventional contract:

- temporary works to adjacent buildings
- access points to the site and any restraints on layout that have to be considered
- existing services; water, electricity, overhead cables, etc
- any security problems
- labour availability
- weather conditions
- necessity for temporary roads
- nature and use of adjacent buildings and services.

LIAISON WITH CONSULTANTS

A visit to the architect would normally be made but it would also be advantageous to visit the engineers and quantity surveyors.

Detailed drawings, reports of site investigations and any other available information must be inspected and notes and sketches made of all matters affecting either construction method or items likely to influence costs.

A well documented project may be indicative of a smooth running and possibly profitable contract. An ill-conceived and poorly documented project may cause delays, and allowances for this should be considered in the tender.

A visit to the consultants will clarify all outstanding queries and ensure that there are no contradictions

6.9 New concrete toe to floor edge with applied damp proof coating. *High rise flats, Birmingham*

between specification and drawings prepared by separate professionals.

Investigations at the consultants offices should include the following:

- an examination of details and working drawings
- an assessment of the degree of complexity of the work
- clarification of any phasing requirements
- resolving queries arising from the project information received with the tender
- discovering whether the consultants have a particular construction method
- enquiring how the contract period (if any stated) was established
- assessing whether the consultants attitude and views suggest an understanding and experience of such projects.

COST FACTORS

COSTING EXAMPLES

Costing and tendering

EXAMPLE 1 *Structural Repairs* (all provisional)

Information Due to the specialised nature of work in this section it has not been measured in accordance with the concrete work section of the 7th edition of the Standard Method of Measurement *Special preamble* The rates for the following items shall include for breaking out damaged or eroding concrete, preparing as described, scrubbing in slurry priming coat as described to pre-dampened repair area; applying cement and sand repair mortar as described to wet slurry; any necessary formwork; finishing repair flush with surrounding surface. *Cut back, prepare and repair concrete surfaces as described; not exceeding 40 mm total depth in one application* General surfaces at eaves A over 0.25 not exceeding 0.50 sm B over 0.10 not exceeding 0.15 sm C over 0.50 not exceeding 0.10 sm	2 3 2	No No No	51 98

Below is how item A of the above bill description is analysed and the rate built up to produce the basic rate.

	S/L	LAB	PLT	MAT	Total
Cut back, prepare and repair concrete surfaces as described. Depth n/ex 40 mm. In one application. General surfaces over 0.25 not exceeding 0.5 m² Break out concrete 3.75 hrs @ £5.00 Clean down prior to priming 0.2 hrs Priming slurry 0.5 m² @ £2.00 Priming 0.2 hrs Repair mortar 0.02 m³ £4.00/m³ Labour to repair 3.5 hrs		 18.75 1.00 1.00 17.50	 1.00 	 8.00 	
Overhead and profit +10% N̲ *Note* All plant and small tools priced in prelims.		38.25		9.00	47.25 4.73 51.98

116

COSTING EXAMPLES

EXAMPLE 2 *Brickwork and blockwork*

	Quantity	Unit	Rate	Amount
Facing bricks; main facing bricks PC £240 per thousand; 215 × 102.5 × 65 mm; in coloured gauged mortar (1:1:6) stretcher bond; weather stuck pointing as work proceeds				
Walls				
(a) half brick thick; facing and pointing one side	9	sq m	29.59	
Skins of hollow walls				
(b) half brick thick; facing and pointing one side	71	sq m	29.59	

Below is the cost breakdown and build up for item (a) on the above bill example.

			S/L	LAB	PLT	MAT	Total
14/A Half brick wall in facings PC £240/1000 m² Coloured mortar							
Coloured mortar 1:1:16							
1.80T Premixed lime/sand	@ £21/T					37.80	
0.27T Cement	@ £55/T					14.85	
						52.65	
Waste in mixing, transporting and laying	+ 20% per m³					10.53	
						63.18	
Bricks P.C.		240.00					
Waste + 10%		24.00					
		£264.00					
H.b. wall 59 bricks/m²						15.58	
Mortar 0.018 m³/m²	@ 63.18					1.14	
Lay 60 bricks/hr	@ 11.00			10.81			
				10.81		16.72	27.53
Overheads and profit	+ 7.5%						2.06
	per m²						29.59

Note
1. Labour for mixing mortar included in laying output
2. All plant included in prelims

117

COST FACTORS

SUMMARY

The particular cost factors which affect refurbishment/repair work, when compared to new build, are:

- restrictions on working in specific arears or to specific elements
- provision of suitable access to the work position
- protection of extra work activities to achieve a repair
- matching new components/materials to existing (colour, type, dimensions)
- allowance for learning curve/training on complex repair activities
- provision of warranties/guarantees/insurances
- supervisory staff for specialist activities
- liaison role if users remain in occupation.

6.10 Cutting back spalling concrete ready for repair. *Chicago, USA*

Some repair activities can be much more costly than new build. A major refurbishment, in gross cost terms, can nearly be as expensive as starting anew. The major savings accrue mainly by completing in a shorter time and/or not having to remove all the building's occupiers. Further analysis on overall costs will be presented in the final chapter.

CHAPTER SEVEN

SITE ORGANISATION

Any repair, maintenance or refurbishment technology must be related to sequence on site. This is vital at the initial specification/design stage as the work inevitably involves a degree of maintaining structural integrity. With new build the contractor does not necessarily have to be told work sequences as there should be no problems with structural safety nor people in occupation. By its nature, refurbishment work is more complicated and can involve many more short sequences. For example, in concrete repair, the sequence is:

- locate defective area (hammer test)
- knock off loose concrete
- cut back to good concrete – clean any steel reinforcement
- agree extent and depth of repair with client's representative
- coat steel reinforcement
- prepare for repair mix
- fix any required formwork
- apply new concrete mix – one or two layers – with curing time between
- allow to cure – strip any formwork.

The specifier must be fully aware of all these stages and clearly describe them, together with any further tests/inspections. This is not possible to carry out in one continous sequence. The contractor for productive and economic working, would want to carry out each of the basic operations for a number of defective areas. Say one level of the scaffold or one satellite platform lift.

OCCUPIED BUILDINGS

In cases where the building is still partially or fully occupied then major restrictions may be placed on the contractor. For example, working only between 8 am and 5 pm, no Saturdays or Sundays; no use of grit blasting equipment as it will penetrate into the building. Here the specification should be specific on what is and is not allowed.

A formal liaison channel should be set up between the contractor and representatives of the occupiers. Prior to work starting a schedule of dilapidation should be drawn up. If a full internal redecoration is to be carried out after repairs then this might not be necessary. An example is given in figure 7.1. This should be agreed by all parties.

Complaints may be forthcoming. These should be formally recorded and dealt with conscientiously. An example of a completed form is given as figure 7.2. It may be that compensation is payable, an example of a suitable form is given as figure 7.3.

If the settlement of a complaint involves a payment of cash or a cheque to an occupier, always obtain their signature on a form similar to the one shown on page 123.

The key to achieving a good liaison channel is the appointment, by the contractor, of a named senior manager. This may not necessarily be the site manager, but they must have the authority to make decisions and act on the contractor's behalf. If this role is required, then it should be stated in the specification and/or contract documents.

SITE ORGANISATION

Dilapidation schedule

Date:
Site:
Attended by:
Weather at time of inspection:

External – Garden

Front fence:
- condition – highlight areas of damage or neglect
  - areas to be removed for access – storage
- gate – including ironmongery

Front hedge:
- general description

Paving and patios
- Surface regularity. Damage. Areas to be taken up – storage

Drainage
- Gullies and surrounds
- Manholes

Planted areas
- General description
- Plants/shrubs to be removed for access

Sheds/Garages
- General description: detail any damage or neglect

Pools
- Note as applicable

External lighting
- Condition – working or not
- Damage
- Cable runs/ducts, etc/switches/voltage

Outbuilding
- General condition
- Type of structure – materials used
- How is it affected by repair
- Services connected – gas, water, electric, BT equipment, meters

Overhead cables

Trees
- Do these obstruct works
- Damage

Existing debris
- General description

External fabric

Front elevation:

Wall
Below DPC – condition, etc
Above DPC – condition, etc

Roof
Damage
Fascia and soffit
Gutter including down pipes
Chimney

Joinery
Windows
Doors
Canopies

General
Side elevation – same as front elevation
Rear elevation – same as front elevation

Internal

BEDROOM 1

Ceiling
Cracks, etc
Existing covering, eg polystyrene tiles
Light fittings
Condition generally

Walls
Cracks, etc
Decoration and condition
Window reveals

Floors
Carpets – condition and approx age, etc

Furniture
General comment

Electrical
Fittings and equipment

Repeat as above throughout the house including meter readings where possible

BATHROOM

Sanitary ware
Condition
Cracks

Plug and chains
Damage
General comment on installation, leaks, drips, etc

Tiling
Condition and type
Number damaged or loose
Condition of grout, etc

NB Include previous items from internal list

KITCHEN

Kitchen units
Condition
Damaged
Fixing especially wall cupboards
Electrical appliances including age and condition

Tiling
Condition and type
Number damaged, cracked or loose
Condition of grout, etc

NB Include previous items from internal list.

7.1 Dilapidation schedule

7.2 A sample entry in a complaints book

Date	Complaint	Nature of complaint
14 Jan 1988 1400 hours 1400 hours	Mr F Jones Flat 28 SMITA TOWER 031-407 1234	Scaffold damaged window pane when dismantling scaffold
Recorded To	**Site Manager Notified**	**Operative arranged Tenant informed**
J Snow C of W J Simmer	15 Jan 1988 1500 hours	Site Manager measured glass and notified tenant that work would be started 10.00 hours 15 Jan 1988. Rang glazer 16.30 and booked repair as above.
Date work carried out and by whom	**Clerk of Works Inspection Date/Time/Comment**	
15 Jan 1988 10.30 hours by Glazing Co Ltd 031-213 6789	Checked okay 15 Jan 1988 12.00 hours Putty to be painted during decoration stage.	

> **WITHOUT PREJUDICE**
>
> Subject to your approval,
> I hereby agree to accept the sum of £   in full and final settlement and discharge of all claims for loss and/or damage occasioned by
> which occurred on or about
>   day of   19
>   Signature
>   Date

7.3 A complaints settlement form

PROGRAMMING

Closely linked to sequences on site is the production of a realistic programme of works. Critical points to take note of are:

- areas available to contractor for work
- time for installation and any subsequent movements of means of access, eg scaffold
- need to carry out safety checks
- allowances for curing of repairs
- allowances for obtaining materials/components specially made to match existing
- need for special components/fixings
- working at heights during winter months.

The means of producing and presenting the programme will be the same as for new build, for example a critical path analysis with presentation by arrow diagram and/or Gantt chart.

INTERPRETATION OF DOCUMENTS

All documents produced at tender stage should clearly describe the extent of the works. Or if a performance specification is used then the contractors are made fully aware of the exact demands of the client.

Contractors new to repair/refurbishment work should be introduced to the requirements by ensuring that:

- all documents are presented in time for the tender period
- the contractors (all on tender list) are interviewed to satisfy the client that they are capable (have the necessary skills/experience/know how) to carry out the work
- if a new contractor wins the bidding stage a close scrutiny should be made of the prices.

This type of work does not lend itself to be carried out by a contractor with only new build experience. If such a contractor is appointed it is likely that close supervision and liaison will be required by the client's representatives. The implications of this statement are that it is very difficult for any new contractor to break into this kind of market. This is true, but it does not preclude their entry. It should be entered with caution and prior knowledge by both contractual parties knowing that a dialogue is essential on all matters. Nowadays, there are enough specialist contractors in the repair/refurbishment market for clients to obtain a competitive tender.

QUALITY ASSURANCE

Some clients are demanding that contractors demonstrate that they have a quality management system. This is not the same as quality control for site activities. Quality control is the enforcement, via tests, inspections and warranties that materials and workmanship is to the required standard. Quality management systems are management procedures which ensure that conformance to requirements are met first time. Quality assurance is the confidence in knowing that the stated objectives are consistently achieved. In order to reflect a quality management attitude, procedures need to be implemented which show how the company ensures conformance to requirements. Although it can be demonstrated through documentation and audit, the real concern

SITE ORGANISATION

Sequence of erection procedures

(a) The first building to be repaired shall be *Solihull* house and before the contract commences tenants will have been removed from 34 dwellings which are particularly forming the south gable end.

It is at this end therefore, that work will commence.

As work approaches completion on this gable tenants in dwellings at the north end will be transferred into the newly repaired dwellings.

The same procedures will follow in turn on the south and north gables of *Leamington* house until the work is completed.

(b) The method of dismantling the defective panels and of the assembly of the new shall be as follows:

1 Scaffolding will be erected on the south/west bay of the south gable of *Solihull* house, to provide a working platform for the removal of defective panels to storeys 1 and 2.

The erection of the new panel to storey 1 will follow immediately and the scaffold will then be extended to provide a working platform for the removal of storey 3 with the erection of the new panel for storey 2 following, and so on in sequence.

2 As the completion of the first south/west bay approaches, work in the same sequence will commence on the south/east bay and in its turn, work will commence on the north/west bay of *Solihull* house and proceed as previously.

In turn, work shall then proceed in the same sequence on *Leamington* house.

3 As the ascending sequences of the installation of the new panels proceeds and the necessity arises for structurally stabilising the scaffolding it shall be the Contractor's responsibility to dismantle the opening portions of certain window assemblies to enable the scaffolders to effect restrain by attaching the framework to the specified cross-wall.

The Contractor shall be responsible for protecting all assemblies from damage during the erection procedures and shall pay the cost at completion of repairs, or total renewal, at the discretion of the Architect.

4 After the defective panels and associated polystyrene sandwich have been removed and before the new insulating layer is applied, the exposed face of the backing-up structural panels, shall be examined and the following work carried out.

(i) All excresences and high points shall be removed in a manner agreed with the Architect and the surface cleaned down and freed from any grease, oil, moss or other lose material.

(ii) The original dry-pack sealings shall then be examined to determine the condition. Where this is found to be missing or displaced, it shall be raked out in stages and replaced with a new packing, all in stages and with materials agreed with the Architect.

(iii) The entire exposed wall shall then be liberally coated with a two-coat application (1 primer, 1 finishing coat) of a bitumen latex damp-proofing emulsion such as MULSEAL DPC or equivalent approved, brush applied to form an impervious membrane over the face of the building, applied strictly in accordance with the manufacturers recommendations and left ready for the further work specified.

Panel sizes, shapes and composition

(a) The panels shall conform to the same external configuration as that established by the defective panels and in respect of horizontal and vertical joints, must align with the original

QUALITY ASSURANCE

> panels which are retained at the point of junction with the returns on to the east/west faces. A schedule of the panels concerned is shown on Drawing No. ...
>
> **(b)** The new panels to the north/south faces shall have an overall thickness at the ribs of 250 mm (against 75 mm for the existing panels) with a web thickness of 75 mm with the breath of rib as shown on Drawing No. ...
> The edge depths will likewise be 250 mm with the joint profiles as shown on Drawing No. ...

is with peoples attitudes and commitment. There not only needs to be a structural (*documentary*) change but a conceptual (*attitudinal*) change in the organisation. ASHFORD (1989) describes, discusses and illustrates the management processes required in order to set up and run a quality management system.

BS 5750 (CEN 9000) provides a standard to which companies can reach. This involves assessment by a third party (perhaps with initial help in setting up the necessary in-company procedures) and yearly auditing of the procedures. A certificate is awarded on the basis of satisfactory implementation of 'getting it right first time' and the company can testify to its claim that quality is assured.

The emphasis is on management – not the physical work process or final product (the repair). The reasoning is that if all the correct information has been clearly conveyed and that all the materials are supplied conforming to requirements, and that the sequence/methodology of work has determined and passed via instructions to the operatives, then it is highly likely that the required quality will be achieved first time. The job should be better planned and organised, materials to hand. This should result in less remedial work to the repairs. It may necessarily involve training in a number of areas. A real example can illustrate this point. The contractor was carrying out strengthening to a high rise large panel system building. This consisted of bolting steel L shaped angle cleats at the internal junctions of the concrete external walls and floors, as shown in figure 7.5.

Work had been progressing satisfactorily for six weeks. Sample pull-out tests by the engineer had shown that the fixing bolts were meeting the required tonques. After a break for Christmas holidays the work continued. The engineer did not test any more for a few weeks. When testing was resumed an alarming number failed. In fact they could be virtually pulled out by hand! An inquest was held – was it:

7.4 Some contract documents will provide clear instructions with regard to sequencing and methods of construction. Replacement of external cladding to Bison blocks. *Portsmouth*

- defective fixing bolts?
- insufficiently weak concrete providing no mechanical bond?
- not tightening up the fixing bolts?
- poor drilling of the holes?
- not driving home the fixings to create the expansive key?

It was found to be the last item in the list. A new gang of operatives had been moved onto the job. Management had failed to explain and train them in the correct sequence of operations and use of tools. If a quality management system had been in place the following would have occurred:

- the site manager would have checked with the operatives that they could do the job
- a form, signed by the manager, would state that they could do the job
- if the operatives could not do the work then the manager would have to train them – this would then be recorded
- the manager (or deputy) would then check the work as it proceeded – without waiting for the client's engineer. A record would be made of the contractor's check.

The quality management system demands a clear definition of roles and responsibilities. Activities should not be left to chance. Any management actions should be recorded. In this case the contractor had to go back and open up all those angles completed since the vacation and check them – all at the contractor's expense. This wiped out virtually all the profit for the

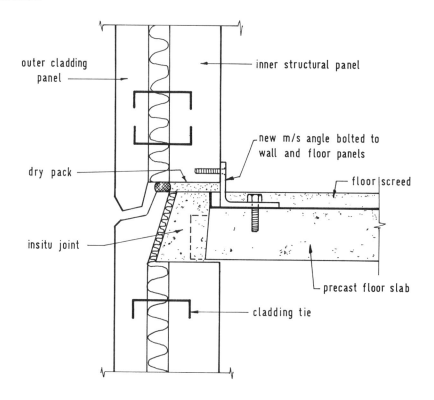

7.5 Angles fixed to LPS buildings as strengthening to resist internal explosive forces. *Bison type*

7.6 Temporary propping to support floors during reinstatement of PRC elements. *Wates houses, Leicestershire*

job – all for the sake of a few hours' interest and actions by the site manager.

The implications of quality management system implementation and certification is widespread. Not only will it affect the contractors but the designers/consultants/engineers will also have to demonstrate that they had systems in place. A major reason for all the present repair and reinstatement work on modern buildings is the past lack of quality achievement by building management and designers. Building owners are wary and do not want to keep remedying further mistakes. In some cases they demand long term warranties for repairs and new components. By instituting a quality management system it is far more likely that quality in all activities will be obtained.

SAFETY

7.7 Specially designed support angles to hold up windows while new brickwork is erected. *College building, Birmingham*

SAFETY

Safety is of paramount importance in the repair and refurbishment of structures. This is the ultimate responsibility of the contractor, although the client's representative will give advice and direction where the structural integrity of the building is at risk.

The main areas of concern are:

- retaining structural integrity of all load bearing elements
- control of loose debris during repairs
- handling of elements being removed
- stability of means of access
- protection to people if in occupation.

Structural integrity is initially determined from an analysis of the structural design. This is then related to the actual members and their condition. The engineer may advise temporary support or brackets to be used to withstand new loads. For example, temporary brackets fixing the inner leaf of a cavity panel wall to the structure. This is to resist the wind suction forces which will apply as the external brick leaf has been taken down and will be rebuilt.

Concrete repairs will produce debris when cutting back to the sound material. This is likely to fly everywhere and precautions must be taken to limit this. Also provision for the collection and disposal of this material is to be controlled.

The removal of external precast panels requires a detailed methodology. Problems can arise in ascertaining the point at which the panel actually becomes detached and therefore held by the removal apparatus. Sudden uncontrolled movements in windy conditions at the top of a high rise building should be avoided. This also applies when new fixings are being installed into existing panels. Some insecurity may be induced if these panel's fixings are inadequate – drilling holes could cause a failure in the original fixings and the panel fall off.

Any items being removed should not be thrown or discarded unless it is via a chute or handled directly into a skip.

The means of access has already been discussed in the previous chapters, but a few points bear emphasis. The stability of all forms of access to high

SITE ORGANISATION

rise structures depends on the structure itself for ties. This is also true for two storey buildings but a scaffold can be virtually independent and can be stabilised externally to allow free access around the walls. In the former case, any ties back to the structure need to be agreed by an engineer as it will involve either drilling into the concrete elements or using other elements as supportive backgrounds. These could put undue local strain on the structure. Constant vigilance must be exercised by the site management team to ensure that no ties are loosened or removed during the works.

When people are still in occupation during the work then their protection is paramount. Temporary protected access and restricted points of entry may be necessary.

EXAMPLES OF SITE METHODOLOGIES AND SEQUENCES

Low rise

Where PRC houses are to be reinstated with load bearing brick/block cavity walls this is carried out under a licensed scheme. This enables the house to be mortgageable. The licence given by PRC Homes Ltd, the authority for their issue, is based on a submission by companies or professional practices wishing to carry out this work. On obtaining a licence they can compete for private and public housing reinstatement work. The submission documents must provide full details of the proposed reinstatement system and show how it will be carried out. The following extracts are taken from a submission for the reinstatement of a Wates house and show the nature of the information required:

'Occupants can remain in the house clearing construction with some inconvenience.'

'In high wind conditions the occupants may have to be temporarily vacated due to a slight risk of damage to the building.'

7.8 Residents can remain in place during some PRC reinstatement schemes. A new party wall being finished. *Wates houses, Nuneaton*

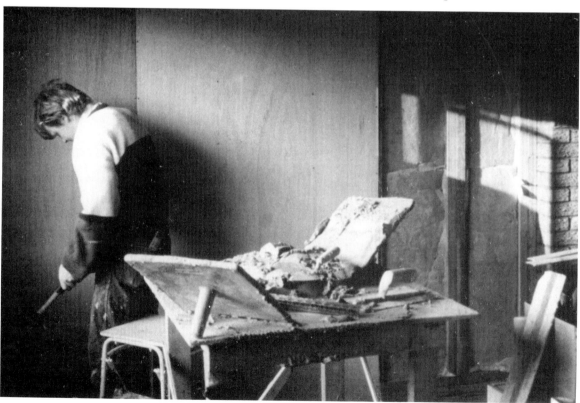

EXAMPLES OF SITE METHODOLOGIES AND SEQUENCES

This is at its greatest when the concrete walls have been taken down in part leaving the eaves exposed which could lead to the roof being lifted.

'Structural temporary screens with windows and doors built in are provided to support the roof and floors, and protect the interior during the building works.'

These screens give support to the 1st floor and roof as well as providing a weather proof (but not thermally good) screen. In winter months these do cause a problem for the occupiers as they do not provide full weather tightness. See figure 7.10.

An alternative method is to use individual support props and lightweight screens. The problem here is that the props can interfere with the normal activities of the householder. For example, in the kitchen they may prevent full access to the kitchen cupboards.

The main stages of a typical sequence for a semi-detached house are shown on figure 7.9. The front and rear walls are demolished and rebuilt with a return on the gable ends. The remaining sections of the gable ends are demolished and reinstated. The party wall is then done last of all.

Even though the sequence is given in the licence document it by no means details the actual work to

7.9 Sequence of work presented in a licensed scheme for PRC reinstatement. *Wates house*

SEMI DETACHED HOUSE

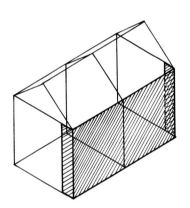

1. Fix two storey screens to front elevation to support 1st Floor and Roof.
   Rebuild front walls with 600mm side returns

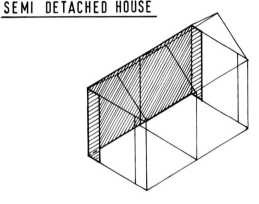

2. Repeat operation 1. to rear elevation

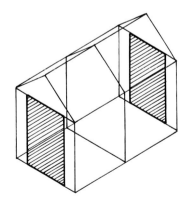

3. Complete side wall, after fixing side screen to both floors, and temporary timber binder to gable studwork

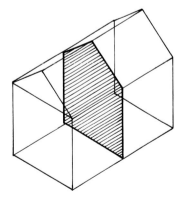

4. Erect two storey screens to each side of party wall.
   Demolish and rebuild party walls

SITE ORGANISATION

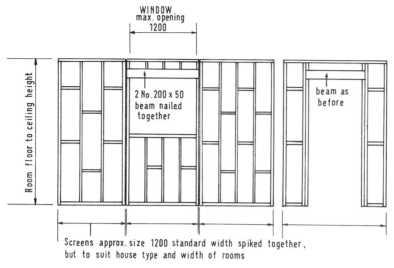

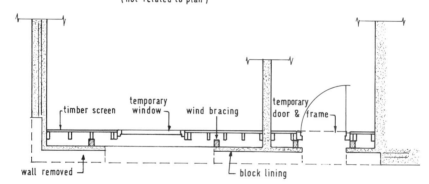

- Screens nailed to each other to form rigid structural panels
- Screens nailed to floors and roof to make structurally stable
- Sole plate of screens to be screwed and plugged to floor slab and end studs screwed and plugged to block lining
- Screens construction as described in 'Schedule of Materials'

7.10 Details showing position and method of construction of temporary screens during reinstatement. Resident remains in occupation.
Wates house

be done and its sequence. The description of the work for the foundations illustrates this point.

'Excavate strip footing around all house, one elevation at a time and drill and epoxy grout 12 mm austenitic stainless steel bars at 300 centres into existing concrete foundations. Place concrete to one elevation at a time, building in continuity bars where joining concrete strips placed at different times.'

In practice the contractor will encounter the services into and out of the building, such as drainage and water supply. Waste pipes may have to be extended to cope with the increased thickness of the new wall. Gullies may have to be moved. In the sequence of operations they will have to be catered for just after excavation for the foundation and before the new concrete is poured.

A further problem may be encountered due to the ground conditions or depth of the existing foundation. If the soil is very poor the building inspector can insist on the new foundation being formed on a good stratum at a lower level than the existing one. This might then involve some underpinning to the existing foundation.

It is only when excavation commences that the actual sequence of work can be finally determined.

High rise

It has already been mentioned in chapter 6 that cost plays a major part in determining methodology and sequence. On large buildings this becomes more important as a matter of scale is apparent. If the

7.11 Temporary access between blocks during construction of new lift tower. *Doddington estate, London*

7.12 Drilling into concrete panels for wall ties as brick skin is brought up from foundations.
Gravelines, France

sequence is not right then extra costs can be incurred as mistakes can be multiplied down the building.

The case of a multistorey block of flats illustrates a typical sequence. This involved the removal and replacement of brick panel walls, undertaking some strengthening, carrying out necessary concrete repairs and completely redecorating internally. The contractor opted for scaffold access as bricks needed to be loaded ready for laying; grit blasting was used to clean the concrete and it provided good working areas.

Work commenced from the top, but only six floors were worked on simultaneously. The basic operations were thus:

- hack off render/mosaic to edges of concrete floors
- clean concrete and reinforcement and carry out repair
- strengthen internal leaf of cavity wall panel
- take down brick panels to three floors – including removal of window frames
- cut off existing wall ties
- drill and insert new wall ties to internal block leaf
- face concrete floor beam with render
- rebuild brick panels and build in existing windows but with new sub frame
- paint and clean external wall
- work to other three floors commencing with concrete repairs and demolition to give continuity for bricklayers for the six floors
- fix stainless steel edge plate
- relocate scaffold boards and sheet protection down to next series of floors.

One or two points should be mentioned which affect methodology. The blocks were unoccupied so the existing lifts could be used. Therefore access to the scaffold was from the inside. This was achieved by taking panels out of the balcony rails. A materials hoist was provided externally.

There was a problem in obtaining properly gauged brickwork to the new external panels. The original building was constructed in imperial dimensions, but the replacement bricks were in metric sizes. The internal leaf remained in place, as did the position (with replacement only of the subframe) of the window. To alter this would have created a great deal of making good to the internal finishes and lintels. The section shown in figure 7.13 illustrates the problem. The initial points are the distances between: floor support to underside of external cill; underside of cill to top of window frame; top of window frame to underside of floor slab. In each section the gauge of the brickwork was slightly different resulting in different thicknesses of the mortar bed joints.

A further complication on this contract was the client deciding late on in the day that internal redecoration was to be done. This caused an overuse of the internal lifts as all personnel, together with light materials and equipment used them for access. Nevertheless, the contract was completed on time.

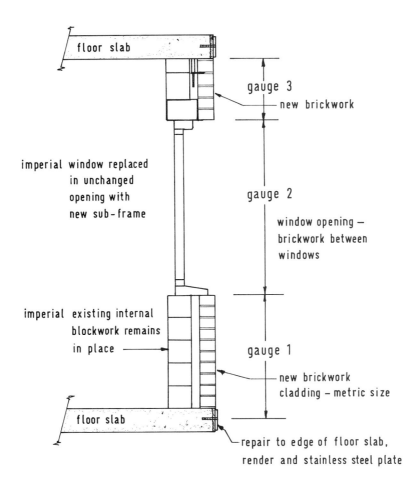

7.13 Dimensional problems in replacing metric sized components into imperial sized space

SUMMARY

Site organisation is the prime responsibility of the contractor, but the specifiers/consultants/client can directly constrain this by their legitimate demands, for example keeping people in occupation or ensuring structural safety. Also the specifier must be fully aware of the details of processes in order that realistic clauses can be written.

Generally an overall contract period is stated by the client, but it is the responsibility of the contractor to derive the actual work programme. This is based on an interpretation of the documents and experience of the work, both related to the particular building.

A natural worry of clients is that any repairs or improvements do not repeat the mistakes of construction. Many defects were caused by the faulty design and construction which is now involving building owners in expensive, and avoidable, repairs. Therefore quality assurance is a paramount requirement. Contractors, together with material suppliers, must ensure that the repair is satisfactory and suitable for the purpose. It must conform to specified requirements, and be achieved at the first time of execution.

A direct responsibility of the contractor is safety. Method statements should be produced covering all aspects, ranging from structural to occupiers to workers.

Repair and refurbishment work is generally far more complex than new build. It is more labour intensive and can involve a wider range of skills, both managerial, technological and craft.

CHAPTER EIGHT

TYPICAL SOLUTIONS

A number of typical solutions embracing basic repairs to major refurbishments will illustrate the range of options available.

Briefly the case studies are:

UK: Housing
One Medium rise, cross wall, precast panel system with brick external cladding; basic repairs. *Walsall.*
Two Concrete repairs, replacement of external brick panel walls and minimal internal refurbishment to high rise block. *Stafford.*
Three Complete refurbishment and structural strengthening to medium rise large panel precast concrete system. *Hull.*
Four Wates houses – reinstatement of external walls and complete refurbishment.

France: Housing
Five Overcladding and extensions to medium rise block of flats, *La Courneuve, Paris.*
Six General refurbishment, structural alterations and overcladding to system blocks. *Savigny-sur-Orge. France.*
Seven Overcladding to 2 storey large panel concrete houses. *Gravelines, France.*

USA: Housing
Eight Apartment building, *Chicago, USA.*

UK: Commercial
Nine Office refurbishment, Edgware Road, *London.*

STUDY ONE

DESCRIPTION

Bison cross wall flats in Walsall, UK. Six storey, large precast concrete panel, load-bearing units with floors spanning PRC cross walls. Gable ends faced with brick leaf, with front and rear elevations having brick/block cavity walls.

Constructed in 1965 – five blocks.

PROBLEMS

The problems were:

1. excessive overhangs at floor support
2. render falling off the edge of floor slabs
3. brick displacement at top of the panel wall
4. erratic cavity dimensions – ranging from 15 mm to 250 mm
5. on large cavity dimension, wall ties placed in concrete panels unable to bridge to brick outer skin
6. unformed soft joints (not included in design)
7. missing wall ties – not placed during construction.

Also it was discovered that the concrete columns and precast concrete edge beams at roof level, creating a clothes drying area, were inadequately fixed. The fixings were also showing signs of corrosion and the concrete was affected by carbonation.

SOLUTION DESIGN

8.1 General view of external works which consist of concrete repairs, brickwork strengthening, floor edge repairs and new pvc double glazed windows.
Bison cross wall, Walsall

SOLUTION DESIGN

A structural engineer was appointed by the building owner who had carried out the original survey on one of the blocks. The state of the brick was generally good. No cracking or spalling but there was displacement of the bricks to the underside of the floor slab. The render finish to the slab edge was also being pushed off due to the compressive forces attributable to concrete creep and lack of movement joints.

The solution was designed to:

- hack off the render and replace with insulation and new render with protective metal covering after painting with carbonation reducing agent
- rebuild brickwork badly displaced
- insert new wall ties to all panels on the gable ends at 450 mm centres
- place angle brackets under the brick overhang at 900 mm centres to provide extra support
- insert fixings tying external brick skin to inner block skin via the edge of the floor slab.

These solutions are shown in figure 8.3.

One block has been repaired and was used to provide information for a detailed specification and construction drawings. On this basis the successful contractor (from competitive tendering) ordered the angle brackets, wall ties and special fixings. Unfortunately the first three blocks undertaken in the five block

TYPICAL SOLUTIONS

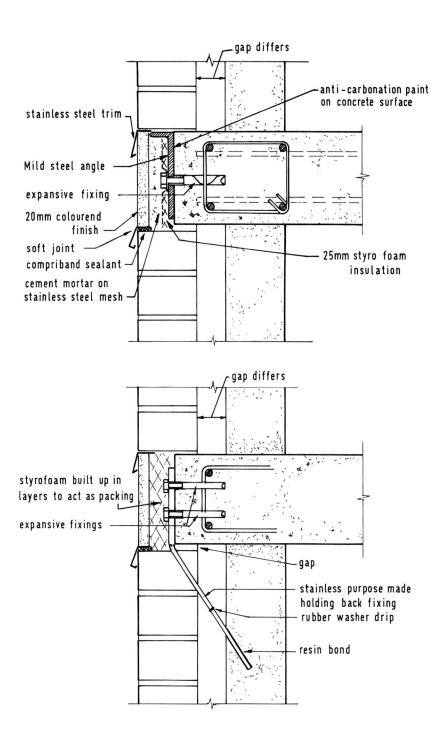

8.2 Repair details. *Bison cross wall, Walsall*

contract did not compare with the first. This was only discovered when the contractor started work in stripping off the existing render to the floor edges and looking into the cavities to insert the new wall ties.

On one floor level the brick overhangs varied between 15 mm and 100 mm. The thickness of existing render also varied from 30 mm to 150 mm. The brick panel on one floor did not line flush with floors either above or below. The cavity gaps between inner and outer leaves varied from 25 mm to nearly 200 mm, again along one length of wall.

The original specification called for one size of support angle cleat, two or three lengths of wall ties and one size of holding back fixing. The floor edge beam was specified to be 25 to 35 mm of cement/sand render with 20 mm of 'colourend' finish, with cover trims.

The main problem was the varying thickness of 'dubbing out' to the floor edge. This could be up to 150 mm, which therefore required a build up of layers. This build up would take extra labour and time, and the contract would far exceed its 44 week period. Also there could be problems with, for example, cracking and delamination, on such a large build up of cement/sand mortar. Therefore a styrofoam insulation sheet was used as packing. In some cases two thicknesses were required and in others the single thickness had to be tapered to suit the changing width of overhang. In order to provide a bond between the mortar and insulation a stainless steel fine mesh was used. Where excessive thicknesses of insulation and mortar were used a further layer of stainless steel mesh was placed to give a good bond for the 'colourend' finish.

Initially the varying brick overhang caused problems in the angle cleat support design. The angle cleats had to be made in three sizes and thicknesses; the longer the support section the thicker the stainless steel. Also the original specification stated 900 mm centres apart. The larger brackets were placed at 600 mm apart to ensure adequate support of the brick panel walls.

The longest standard new wall ties (375 mm) needed to be used to cope with the excessive cavity gaps. The operatives had to measure each hole depth from the penetration of the drill and select the right length wall tie. As would be expected, the operatives needed time to learn before meeting productivity targets. This was also so for the fixing of the angle cleat supports.

The purpose-made holding back fixings to the top of the brick panels gave the least problems. In some cases the drilled bricks cracked as the hole was too near the arris. Some problems were encountered when drilling into the floor edge when the reinforcement bars were hit. This also affected the fixing of the angle cleats. When full penetration was not possible a resin anchor was used instead of the expansive type.

A full scaffold was erected around each block with material hoist. Operatives used a ladder access. The sequence of operations started from the top and worked down, although the work to the concrete columns on the roof was finished last.

8.3 Support brackets to upper brick wall fixed to edge beam. *Bison cross wall, Walsall*

TYPICAL SOLUTIONS

At the same time the original timber framed windows (now quite rotten) were replaced with double glazed Upvc units.

COSTS

The original contract sum was for £1.7 million, but owing to the unforeseen condition of the blocks (necessitating the extra angle cleats, work to the floor edges and wall ties) rose to £2.0 million, a near 18% increase in costs. The contract was completed in the required 44 weeks. This works out at approximately £8,300 per flat.

EXPECTATIONS

The public housing client wants as long a life as possible from these flats. The coating for reducing the rate of carbonation is warranted for 20 years. The engineers foresee no further structural problems and therefore a life exceeding the flats original 60 year expectation is not unrealistic.

The most extensive brick repairs occurred on the corners of the gable ends.

As previously mentioned major repairs had to be carried out to the concrete toes. This involved use of formwork to create new moulds. The curing of the rebate for the soft joint by disc cutter was difficult as working space was severely restricted by the existing brickwork. No allowance had been made to take off the top course of bricks to give adequate working space. This lack of space also caused problems for the reinstatement of the concrete toe.

The contractor decided to train basic site operatives in the skills of installing the new anchors and ties and to carry out class 'B' concrete repairs. The class 'A' repairs, using proprietary materials, was carried out by a specialist sub-contractor. Brick replacement was carried out by bricklayers and also the operatives who had inserted the ties. The new learning and transfer of skills was carried out on the job under the instruction of the tie manufacturers, supervising engineer, clerk of works and site manager. The site manager had no previous detailed experience of this type of work. Progress was slow in the early days, but quality was good. Productivity did increase substantially.

The residents remained in place during all the works.

STUDY TWO

DESCRIPTION

Two 16 storey blocks of flats in Stafford comprising 91 flats in each. In situ concrete frame with in situ concrete floors. External walls in facing bricks. Internal leaf to walls in hollow blocks laid in cement mortar, tied together with butterfly wall ties. Steel windows in timber subframes. Concrete floor edges rendered and faced with mosaic.

PROBLEMS

Deflection and cracking was occurring in the brickwork with concave bowing. This was particularly apparent under the windows where vertical cracks followed down the line of jambs. Cracks were also at ground level adjacent to the corners and the damp proof course. The stability of the brick panels to the front and rear elevations was doubtful as the cracking was quite severe. The engineer had already investigated a similar block of flats and had found it necessary to take down and replace the bricks. The lessons learnt from the first block were translated onto the other blocks and it was decided to remove and replace the brickwork rather than just retie the panels. This would give a higher confidence rating for the future life of the building. The client wanted no further external maintenance to the building for another 30 years, other than redecoration. The wall ties were found to be corroding.

The specification, bills of quantities and drawings were based on the experience of the first block. A different contractor won the tender for the subsequent two blocks.

SOLUTION DESIGN

8.4 Concrete repairs, replacement of brick cladding and general internal redecorations to 16 storey block of flats. *In situ concrete frame, Stafford*

8.5 Dry lining and plaster skin coat to make good inner leaf. *In situ concrete frame, Stafford*

SOLUTION DESIGN

As the worst cracking was confined to the panels below the windows on the front and rear elevations, these elevations had their brickwork removed and replaced. The brickwork to the gable end was to be retied. The existing metal windows were reused with a new timber sub-frame. The floor edge facing was to be stripped off, rendered with a polymer modified mortar. A flat stainless steel plate was bolted through the mortar into the slab edge using resin anchor bolts. This gave a 'finish' to the line of the floor.

Each block was empty for the works. A full scaffold was erected and five, then six, floors were worked on at once. Protective sheeting was used which necessitated restraining bolts and scaffold ties to be specially designed. The scaffold was tied into the in situ concrete frame. A small problem was that not enough space had been allowed for the tightening and subsequently removal of these brackets and three or four bricks had to be left out to enable access. This meant more than expected making good to the brick facing walls to hide the positions of the scaffold ties.

At roof level a high parapet was faced in mosaic. This was repaired. A problem was in obtaining the correct match of mosaics together with using an effective adhesive. Delays were incurred in resolving the colour match.

TYPICAL SOLUTIONS

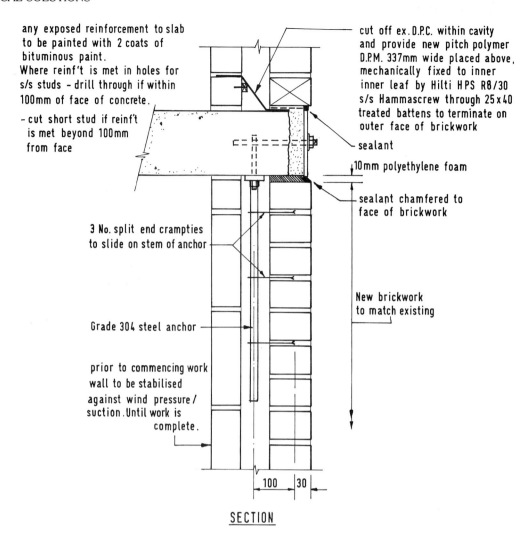

SECTION

SEQUENCE OF WORK

- the inner leaf of the cavity wall was supported with bracing and brackets to withstand the air pressures after removal of the external brick leaf
- the floor edge render and mosaic was removed
- the concrete edge was reformed where necessary with a standard concrete repair
- existing brickwork taken down
- new cavity trays installed
- old wall ties cut off
- new mini-ties to the outer to inner leaf
- rebuild brick panels in new bricks, creating soft joint at the head
- build in new sub-frames and windows
- fix sliding vertical ties to under-side of concrete

8.6 Stainless steel ties fixed to underside of floor slab providing shaft for sliding movement of wall ties holding new brick cladding. *In situ concrete frame, Stafford*

floor slab to restrain the panels against wind pressure and allow for some movement
- fix stainless plate to floor edge
- insert flexible sealants at all joints.

There were some minor problems that required prior consideration. As the internal leaf blocks were hollow a special tie had to be designed. Also on the gables a strap was placed on the internal (room) face of the wall to act as a distribution anchor. This is shown in figure 8.8. This compensates for the relative

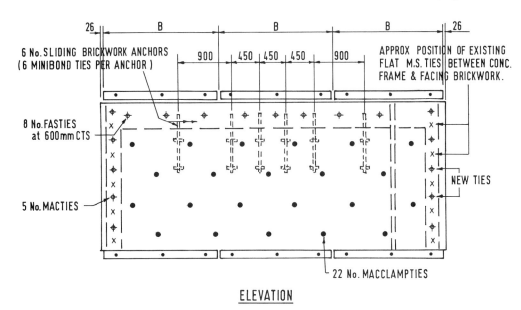

8.7 Internal elevation of gable walls showing position of ties and flat bar spreaders to strengthen the brick claddings. *In situ concrete frame, Stafford*

8.8 Straps distributing load from wall ties. Dry lining will be used to hide straps. *In situ concrete frame, Stafford*

lack of fixing into the hollow blocks. As the gable brick panels were not taken down and rebuilt the sliding vertical ties had to be inserted from inside. A hole had to be made in the block wall and ties anchored into the back of the existing brick wall and the main vertical tie up to the soffits of the floor. An internal wall lining hid these holes and the spread anchor straps.

The new brick walls were coursed and gauged in metric dimensions – the storey heights and window frames in imperial feet and inches. This meant that to put the windows back into their same positions the thickness of mortar joints below and above the openings had to be different. This allowed for the dimensional variation between metric and imperial to be accommodated and disguised.

A variation had to be made on the second block as a shortage of new bricks meant that the window panels had to remain in place. But as bricks were required to replace badly cracked/spalled bricks, the ground floor panels were removed and the sound bricks cleaned and saved. The limited number of new bricks were used only on the ground floor. The saved bricks were used sparingly as necessary.

COSTS

The contract sum was for £950,000 for the two blocks but this included internal decorations. Approximately £800,000 was for the structural works. With 91 flats in each block this produced an average cost of £4,400. The contract time was 65 weeks and it was completed in 65 weeks. The contractor made the estimated profit.

The liaison and information flow between the engineer and contractor went well. During the work the contractor had to change a specialist sub-contractor on the concrete repairs as the quality was not being achieved without supervisory harassment and redoing work.

STUDY THREE

FOUR STOREY, LARGE PANEL SYSTEM MAISONETTES, HULL

Two estates:
Bransholme 627 dwellings now all demolished.
Thornton 558 dwellings. Three blocks demolished, others strengthened (except three) and one refurbished.

In 1984 it was decided to carry out a pilot refurbishment to one six storey block.

There is a strong tenants group on the estate and they have been continuously involved with the design and build process. They were very much part of the process. There are still regular monthly meetings.

The design took about 18 months. In 1984 there was very little in the way of precedence for this type of work and the architects had to proceed cautiously. Design details had to be carefully produced and much discussion was undertaken with materials suppliers.

The existing form and layout of accommodation was to remain. The following concepts were incorporated into the design:

- defensible space
- change of image
- good landscaping
- cheap to run.

The main elements of the works are:

- insulation and rendering to the external walls
- strengthening to the gable walls with concrete buttresses
- new pitched roof and overlay insulation to flat roof
- replace windows with s/w and h/w sills frames using existing double glazing, new entrance porch and halls
- reposition refuse chute
- cover-in access balcony to upper floor maisonettes
- carpet access area on balcony onto boarding
- clad stairs, lift shaft and gables in brick skin
- mini district heating system feeding water-filled radiators
- minimum continuous heat from exposed pipes in each flat

FOUR STOREY, LARGE PANEL SYSTEM MAISONETTES

8.9 Rear elevation of refurbishment to four storey YDG type maisonettes. *Hull*

TYPICAL SOLUTIONS

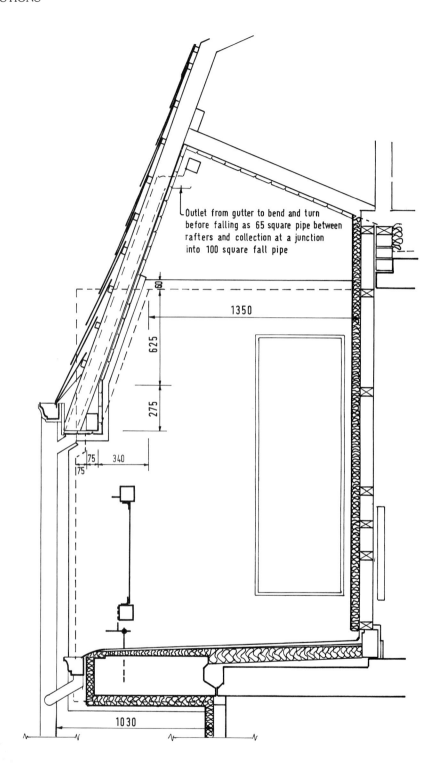

8.10 New pitched roof over top balconies. *YDG, Hull*

FOUR STOREY, LARGE PANEL SYSTEM MAISONETTES

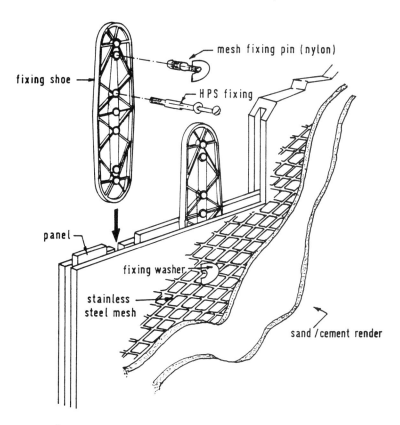

- use of prepaid meter cards
- trade off between new kitchen units and redecorations – tenants choose kitchen units
- new external drain pipes to take water away from building – private balconies now slope out
- new fire escape staircases in steel at gable end opposite entrance controlled entry system
- some concrete repairs to spalled panel edges and corners.

DESCRIPTION

100 mm stryrofoam insulation pinned directly to walls. Epsicon system to general walls carried on polypropylene shoe with expanding bolt fixings. Stainless steel mesh fixed through insulation to hold render. Originally one coat deemed to be all that was necessary by supplier but architect stated it did not cover as it began to crack on drying out. Used scratch coat and then finish all at expense of supplier. Insulation to top and bottom of balconies. Removal of concrete balustrades.

Contract value was £800,000.

8.11 Details of external insulation and overcladding. *YDG, Hull*

PROBLEMS

The existing expansion joints were not fully continued in the new cladding. In the area which is now enclosed on the access balconies and lift/stairs cracking is occurring in the rendering. Mainly at the joint between the support beam and ceiling lines along the walls. Also some slight movement in a panel adjacent to a door. All probably caused by the concrete drying out and/or being affected by the higher temperatures as now protected from the elements.

Some difficulty in producing vertical brickwork to the buttresses in later blocks. Reason not known but could be poor workmanship or not allowing for any misalignment in the existing panels or new concrete walls.

TYPICAL SOLUTIONS

8.12 Enclosed entrance to all floors. *YDG, Hull*

STUDY FOUR

WATES HOUSE REINSTATEMENT

There are over 20,000 Wates houses in the UK (BRE Report 1983). Under the 1985 Housing act private owners, subject to certain conditions, can obtain a government grant of 90% towards the costs of replacing the defective concrete structural elements with a new load bearing wall. Normally this is a brick/block insulated cavity wall, but could be timber frame with brick cladding. In order to qualify for this grant the reinstatement scheme is to be licensed. A number of schemes have been approved and the one described entails the temporary displacement of the inhabitants.

SEQUENCE OF WORK

- all fixtures and fittings are removed and stored after disconnecting the services
- some demolition of external wall panels can commence with the removal of window and door frames
- propping of both first floor and roof commences
- removal of the main support beams to first floor. These can be difficult to take down
- demolition of the party wall concrete elements
- construction of new party wall
- removal of sections of the external wall, commencing from rear elevation
- rebuilding new brick/block walls off existing foundation, commencing at rear elevation, building on new windows and door frames
- when rear wall section complete, carry out removal of PRC and build new wall
- repeat operations to gable ends
- settle structure onto new walls at each floor level. Remove props
- carry out internal plastering after relocation of services
- make good to ceilings, etc
- replace fixtures and fittings, redecorate.

8.13 Removal of party wall. *Wates houses, Leicestershire*

COSTS

The overall cycle of operations is normally carried out in a 10 to 12 week period. Costs, in 1990, are in the order of £23–25,000 per dwelling. This depends on the number of houses in a contract. In some instances local authority owners also carry out a reinstatement at the same time as the private owners. This is usually dependent on finance being available. It is usually the case that in a row or a pair of semi-detached mixed ownership prevails. Therefore when a local authority does not carry out a reinstatement the situation arises of joined structures being in two different materials. This is an example of external factors affecting the technological solution. As it is necessary to carry out reinstatement to a Wates house because of defects affecting the structure, then the same technological solution ought to be applied whatever the status of ownership. There is now a possibility of future problems when each house performs differently in the environment. On economic grounds it is far better to carry out reinstatement now rather than defer it. (CHANDLER 1990.)

TYPICAL SOLUTIONS

8.14 Stainless steel connection ties resin bonded into existing foundation to accept new foundation. *Wates houses, Liecestershire*

8.15 Bringing up external load bearing brick/block cavity wall. *Wates houses, Leicestershire*

STUDY FIVE

LES FENETRES DE BALZAC

Balzac is a long 16 storey block of 320 flats in La Courneuve a suburb of Paris.

A similar block adjacent to Balzac has been demolished as the structural and fabric repairs were considered to be too expensive.

A number of problems were found in this large panel system building. Joints between panels were defective, some concrete was deteriorating, insulation was poor, public areas were unattractive. In addition the building was orientated on an East-West access. This massive slab overshadowed flats, shops and offices to its north. Therefore to improve environmental aspects three large holes (les fenetres) are cut through the building to provide directed light onto the shadowed buildings.

The major construction works consisted of:
- forming the three windows through the building
- altering and improving the ground floor entrances (involves alterations to flats)
- overcladding the gable ends
- general repairs, redecorations externally
- internal improvement of kitchens and bathrooms
- replacement Upvc windows
- strengthening to panels to withstand 10 tonne outward force.

8.16 Overcladding to gable ends, punching large holes through building and general refurbishment. *La Courneuve, Paris*

TYPICAL SOLUTIONS

8.17 Repairs to concrete floor edge beam. *La Courneuve, Paris*

8.18 Extensions to flats and new large 'window' openings. *La Courneuve, Paris*

8.19 Removal of external panels and new walls reshape ground floor entrance to flats. *La Courneuve, Paris*

From the beginning all tenants were involved in guiding decisions after the main concepts had been laid down. The client provided a permanent liaison officer who dealt primarily with the tenants own issues during the construction work.

STUDY SIX

RESIDENCE CHATEAUBRIAND

These three blocks of 160 flats are five storey precast concrete panels blocks built in a rectangular plan shape. Each flat spans from front to rear elevation. The accommodation available was deemed to be on the small side especially the bathroom. The concrete external walls were showing signs of disrepair and had a low thermal insulation rating. The district heating system could just cope with the demand, but the tenants had to pay relatively high bills.

The work consisted of:
- extensions to the bathrooms and living rooms
- external insulation and rendering to all the walls
- insulation to the ceiling of the underground storage areas
- renewal of the sanitary fittings
- installation of new heaters to the bathroom, overhaul of the common heating sysem (underfloor hot water pipes)
- new balconies
- new double glazed Upvc windows
- upgrading of the common entrances and security system
- overhauling the electric wiring and equipment

The significant feature about this upgrading is the extensions to the living and bathrooms. These consist of prefabricated timber units, erected complete with walls, floors and windows. The existing window openings were enlarged after placing each extension one upon the other.

In ascertaining the problems and requirements for these buildings two studies were undertaken.

1 Preparational study which comprised:
- energy audit
- list of actions to give further life
- replanning of flats to seek to gain in habitable area, especially to bathrooms and living rooms
- insulation to the envelope
- upgrading the central heating
- upgrading the sanitary goods
- upgrading the security system
- an architectural quest for a better image of the blocks, with a much stronger integration into the adjacent town centre which was in the course of re-construction and renovation.

2 Sociological study which comprised:
- a survey of the expectations and opinions of the tenants
- the functionalism of the flats
- considering the evidence relating to each set of occupants to review the degree of obsolescence of the fixtures and fittings
- the tenants aspirations to the value of the image of their residence.

Throughout the tenants were consulted, individually and collectively.

COSTS

The costs averaged £15,000 per flat (1988 prices).

Rents rose by about 40% after the works. On average each flat gained 32% extra habitable floor area. The rents are made up from three items; principal rent, general charges and heating charges. The principal rent increased dramatically (this reflects the value of the flat, based on construction costs, loans, land value and loan charges). The general charges remained about the same (for general maintenance and upkeep and management). The heating charges decreased by around 45%. This reflects a major increase in thermal efficiency, primarily gained

TYPICAL SOLUTIONS

by the external wall insulation. As gross flat volume was increased this gain could have been greater if the extensions had not been included.

8.20　New balconies and walls insulated and rendered with polymer cement. *Chateaubriand, Savigny-Sur-Orge*

8.21　Bathroom extensions and timber framing decoration. *Chateaubriand, Savigny-Sur-Orge*

STUDY SEVEN

HOUSES AT GRAVELINES, PAS DU NORD

An estate of large panel concrete houses, built in the mid 1970s was showing signs of concrete deterioration. Also large heat losses were occurring due to the poor thermal efficiency of the walls. Their appearance was unattractive. The scope of the work was primarily confined to the exterior, with some inernal work:

- new foundation and facing brick leaf to ground floor
- prefabricated timber stud framework to upper floor, with vertical tile cladding
- upgrading of internal sanitary fittings where necessary.

The ground floor brickwork was supported on a reinforced strip concrete foundation. Stainless steel thin rod ties were used to connect the new brickwork to the existing concrete panel. A hole was drilled in the brickwork to the existing concrete panel. A hole was drilled in the brickwork and the tie pushed home into a plastic wall plug as the work proceeded. The corners of the brick walls were reinforced in the bed joints.

The upper floor cladding was prefabricated off site and lifted into position by hand. This contrasts with the use of small mobile tower crane for the unloading and distribution of the facing brick packs. The timber panels sat on the new brickwork and were fixed back to the concrete with mechanical expanding bolts. Finally the vertical tiling was fixed after the erection of an access scaffold.

Internally the work was confined to replacing damaged or unsatisfactory fixtures and fittings. In some cases new screeds to the ground floors were

8.22 General view of overcladding to concrete panel houses. *Gravelines, France*

TYPICAL SOLUTIONS

8.23 Fixing clay tiles to upper storey of houses. *Gravelines, France*

necessary as the originals were showing distress. Houses were vacated for these operations, but the majority of tenants remained in place.

COSTS

The average cost (1988 prices) was £15,000.

The main benefits derived from the work was a major improvement in appearance and thermal efficiency, together with a reduction in the rate of concrete deterioration.

STUDY EIGHT

APARTMENTS, CHICAGO, USA

The building is constructed in in situ concrete frame, concrete floors and brick cladding. It has a single boiler unit serving the individual apartments with heating and hot water.

The apartments suffered from inadequate heating, condensation and out-dated and defective sanitary fittings. In addition there were no facilities for laundering and security in and around the block was a problem.

The structure was basically sound. Some repairs were required to the external brickwork. The concrete frame was not showing any signs of distress. The

Chicago Housing Authority decided to upgrade the building and to provide some communal facilities and better security.

The work was:

- repair defective brickwork
- replace windows with double glazed units
- renew all plumbing, electrical installations
- install new kitchen and bathroom fittings
- install new heating emitters
- provide new boiler system, with stand-by – this to have one boiler for heating/hot water and one for hot water only during the summer
- new lifts installed
- on the ground floor communal laundry and drying area
- high security grilles to ground floor facilities

8.24 Fifteen storey in situ concrete and brick cladding. *Apartment building, Chicago, USA*

TYPICAL SOLUTIONS

APARTMENTS, CHICAGO

8.25 New opening, double glazed windows and security grilles. *Chicago, USA*

8.26 New heat emitters from energy efficient boilers giving background heating. *Chicago, USA*

- general improvements to common access areas and complete redecoration to all units.

No problems were encountered during the refurbishment until the latter stages. The paint to the ceilings in a number of apartments peeled off soon after application. The direct cause was not ascertained but was probably a combination of high humidity (condensation) and chemical reaction between the old paint and the new. The ceilings were having to be completely stripped and cleaned and then repainted.

STUDY NINE

OFFICE REFURBISHMENT, EDGWARE ROAD, LONDON

The building is a typical 1960s 16 storey in situ concrete framed structure. The floors are in situ concrete solid slabs. The original cladding was bolted to frame and floors. The building was in good order structurally and due to an existing long lease still remaining on the lower part of the site caused the owners to favour refurbishment rather than demolition and redevelopment. This would have taken at least six months longer, and of course displaced the tenant. The existing floor-to-ceiling heights were adequate to accommodate current service requirements, therefore a full access raised floor system was

8.27 Stairs repositioned on gable end and overcladding existing brick skin. *Edgware Road, London*

TYPICAL SOLUTIONS

8.28 Installing fan coil units in extended floor areas. *Edgware Road, London*

8.29 Refurbishment to commercial buildings. *Edgware Road, London*

installed. A new suspended ceiling incorporating lighting and air extracts allowed an air conditioning system to be fitted. This was made more practical by the 300 mm extension of the floors. A simple gallows bracket bolted to the structure created further floor space around the building. This enabled the fan coil units to be placed outside the line of columns, thereby not infringing upon the lettable floor space.

Also the internal stairs were repositioned onto a gable end of the tallest block, creating further internal lettable space.

The gallows brackets supported the new cladding, a lightweight assembly of fire resistant dry-lining boards with tinted glass external skin. Full height oriel windows gave further floor area and helped to conceal large vertical air ducts which went to additional air handling plant on the roof.

The brick clad gable ends were overclad with powder coated aluminium panels, supported on aluminium mullions with an insulation sandwich.

COSTS

Above the lower block a new floor has been added using a steel frame. In total a further 25% of usable floor space has been created. This helps to provide extra revenue to fund the improvements and refurbishment to all services, including the lifts. It was estimated that redevelopment costs would have exceeded the refurbishment costs of £1500/m^2 by an extra £400/m^2.

8.30 Section through floors showing floor extension and services accommodation. *Edgware Road, London*

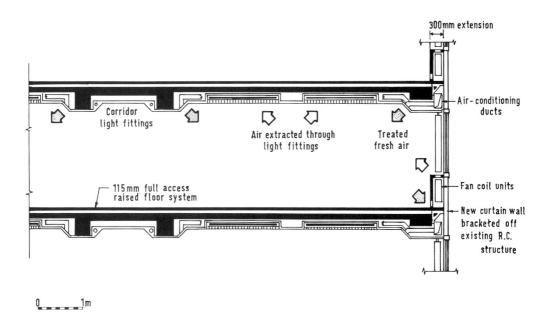

CHAPTER NINE

IS IT WORTH IT?

All the preceding chapters have taken it as axiomatic that refurbishment, or the lesser options of repairs and maintenance, are worth it. But there are many arguments for demolition and total redevelopment which are sound and can be applied to a relatively new building. An example already cited is a 1960s shopping centre being demolished and the site, together with adjoining land, redeveped into a larger complex. This required the diversion of a river and expensive ground stabilisation before construction could begin. The developers took the view that the existing buildings were inadequate for profitable business after refurbishment. A large capital investment in new build would bring greater monetary gain and provide better facilities for shoppers.

Demolition could also apply to housing, although this does have greater social implications in that people have to be rehoused. Nevertheless, the situation does arise where refurbishment cannot be effective because the basic structure is unsound – it has run its functional life. This applies particularly to some types of precast concrete houses, or those built in concrete blocks. A case is the Boswell type houses constructed during the 1930s in the Midlands of the UK. The walls are constructed in two skins of concrete blocks, tied together and built between precast concrete corner columns. The external face is rendered to give a weatherproof membrane. The problem here is that the concrete blocks were manufactured using colliery waste as an aggregate. This has reacted with the cement binder and the blocks are turning into dust. This coupled with carbonation of the columns makes the houses structurally unsound. A reinstatement system could be designed, but the works would be so extensive that it becomes a rebuild. The occupants would have to be temporarily rehoused. Another factor pertinent to the Boswells are that they were built in large groups in estates. Gardens and public space was generously provided. If they were demolished the planning gain could be achieved. Where there are now eight houses, ten could be built. This provides sorely needed houses in a county with little enough suitable housing sites and would generate extra moneys to pay for the redevelopment. In this case one Midlands local authority owner, together with the private house owners on the estate, have agreed that redevelopment is the best option.

In assessing whether it is worth refurbishing, a number of factors need to be considered and then measured against each other. Some of these factors have already been mentioned in chapter 4, Options. Now is the place to take a wider view and see some comparative measurements. The factors are:

- functional life remaining
- need and efficient use of building space
- technological constraints on refurbishment
- degree of disruption to building's occupants
- retention of community – in housing
- retention of original energy units used in construction
- cost of energy units in refurbishment or new build
- environment around the building, such as car parking, services, architectural perceptions
- planning gain possible on redevelopment
- usable space gain to original floor layouts
- ability to achieve present day amenity standards
- planning/legislative constraints
- time for on-site works
- appearance and architectural perception
- cost –comparative to new build
    - capital required
    - sources of finance
    - recovering of costs

COST EVALUATION

9.1 New balconies to flats but roofs give too much shade to rooms necessitating change in design.
Doddington Estate, London

– measure of profitability
– social benefit.

It is the integration and formulation of a matrix of these factors which will produce a solution that is the optimum at the time and for the building in its context.

It would be very useful if the factors could all be brought down to a common numerical measurement. Each factor would be assessed with respect to the building and an evaluation of its worth stated. Positive and negative values might be one way of comparing demolition and refurbishment. Then a rating and cost figure developed, such as the system described in value engineering systems, to give sum totals for different repair/refurbishment options.

COST EVALUATION

There are some models which could be used for evaluating options. WARN (1980) has put forward a working model for the choice of modernisation measures. The hypothesis is that by combining a set of typical conditions before modernisation, with desired conditions after modernisation, which takes into account costs, legislation and regulations, then an optimum solution should be possible. This method is termed the Rational Conversion (RC) Method.

Another model with a definite mathematical equation is presented by NIERSTRASZ (1980). This allows a comparison to be made between modernisation and demolition of the building with no planning gain. It compares the rent after modernisation with the rent after demolition and construction of new dwelling. The formula given as:

$(a_t + m)(1 + y)B < a_{50}(1 + s)B + (a_{50} + m)fB$.

Where:

a_t is the annuity factor expressed as $\dfrac{i}{1 - (1 + i)^{-t}}$ (i is the rate of interest)

t is the period of depreciation

a_{50} is the annuity factor with depreciation of 50 years (or any other set period)

IS IT WORTH IT?

m   is the annual expense on maintenance and taxes, etc, as part of value (B) of existing structure
y   is cost of modernisation as part of value (B) of existing structure
B   is the value of existing structure before modernisation
s   is the cost of demolition expressed in B
f   is the cost of construction of new building expressed in value (B) of existing structure.

This formula relates the factors in terms of B, that is the value of the building before modernisation. The value of the site is disregarded from both sides of the equation.

Application of this formula has been utilised in evaluating the economics of the reinstatement of PRC houses (CHANDLER 1990). The case is argued that of the four options of minimal repair and protective paint; overcladding with insulation and render; structural reinstatement in load bearing brick cavity wall; or demolition and rebuild then, on the same site, reinstatment is the best overall economic solution for all parties.

The principles of NIERSTRASZ'S thesis could be equally applied to commercial buildings of most categories.

Returning to the list of factors each one will have a greater or lesser influence according to:

- form of ownership; (private, public, landlord, investment)
- use of building; (housing, factory, commercial, leisure).

For example, an investment owner of an office building will look to the maximisation of money over a limited period of time. This will be coupled to whether it will be rented out or sold on. If rented out then the rents will be set for a period after which reviews will take place. This will enable future profits to be forecast and income generated at a steady rate. If sold then a larger real (and possibly percentage) profit may be desired in order to create new money for further developments.

In the case of public sector housing different criteria can take precedence, such as demand for housing in the area, the habitability (with regard to health and comfort) of the dwellings and the ability of people to pay for the rehabilitation works. Until recently the full cost of these works has not been passed onto the tenants. Changing accounting procedures will place the burden of any improvements, maintenance, etc, onto the housing revenue accounts. This means that all works will have to be paid for from income arising from letting the properties.

In France the payback period for improvement works on social housing is 15 years. This means that loan monies must be repaid, with interest, by the tenants. As was seen by the case study on the Chateaubriand blocks of flats, the rent increased by 40%, but in this case was mitigated by an increase of 32% of habitable floor area.

The concept of helping to pay for refurbishment by increasing lettable (or saleable) floor areas is not new. In buildings constructed in load bearing brickwork, with façades of architectural interest which have to be retained, the developer will try to place more floors behind this façade. Generally a new framed structure will be built after demolition of the heart of the building. This will enable floor to ceiling heights to be reduced (if previously generous) and/or form extra floors at the top and in the basement.

When considering extra space in framed or large panel buildings this becomes difficult as the work has to be designed around the structure which is generally inviolate. One method now being adopted, with ground space allowing, is to extend floors and then fix a new cladding. One such office refurbishment in London termed by the architects as a rebuild, has increased its 12 m deep width to 15 m by extending 1.5 m either side. This adds a further 22% of usable space at each floor. These projections provide room to house the bulky equipment used in modern air conditioning systems. The structural frame was analysed to ensure it could take the extra loadings. Together with an extra floor at roof level the building took extra wind loadings. Stiffening to the existing concrete lift core walls provided the necessary resistance (SPRING 1989).

The refurbishment budget of £1200/m² is close to the costs of new build. But the construction time for the refurbishment is 17 months compared to about

9.2 Stone cladding of inadequate thickness to cope with climatic conditions being replaced. *Offices, Chicago, USA*

9.3 Damaged and misaligned panels during construction. *Prague, Czechoslovakia*

IS IT WORTH IT?

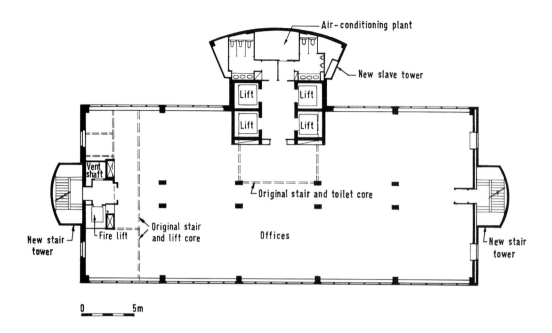

9.4 An increase in floor area helps to pay for refurbishment. *Multistorey offices, London*

36 months for redevelopment. This provides a 'bonus' of 18 months rental income. Additionally, planning control may not allow a new high building. Also the present basement car parking capacity would be reduced in order to restrict road traffic in London.

ENERGY EVALUATION

Another factor now being brought into the equation is energy use. There is a two part argument. On the one hand is the use of energy in itself, that is the depletion of resources. Most energy sources currently exploited are finite – they have a limited life. Renewable sources are not yet commercially and safely developed. Therefore reliance is still on fossil fuels. There is emerging concern of 'global warming'. The use of energy and suspect chemicals in manufacturing and so on is considered to be changing the earth's atmospheric balance. Although no clear indications have been found as to cause and effect there is evidence to show that CFCs might have an affect on the earth's ozone layer. Putting these two aspects together makes the recycling of buildings a much more 'earth friendly' activity. By retaining the basic structures there is no need to dig more aggregates, mine more iron ore, or manufacture more cement, this produces an energy saving plus. Additionally (although equally true of new build) most refurbishment schemes incorporate current thermal insulation requirements and utilise efficient heating and service equipment. New products and components used as claddings, etc, can be manufacturered using inert chemicals and materials. Some developers in London are asking architects and engineers to produce a 'green rating' to their proposals. A points system will be used to compare schemes on the basis of the adoption of energy and ozone friendly materials and construction. These measures will give added weight to the advisory codes produced by the Chartered Institute of Building Services Engineers. The four codes give advice and methods of calculating energy usage in existing and proposed buildings (CIBSE Codes on Energy 1983.).

SOCIAL CONCERNS

A major factor determining the options for the refurbishment of housing are the views of the occupants. We now, rather belatedly, have recognised the need to involve occupiers in devising rehabilitation

SOCIAL CONCERNS

9.5 Broken and out of plumb gable end panels to high rise flats. *Prague, Czechoslovakia*

schemes. This is especially applicable in the public housing sector. Mention has been made of the work by COLEMAN and POWER and the need to evaluate their concepts with respect to layout and management of modern housing estates.

The good intentions of providing decent housing quickly and efficiently and moving people from so called 'slum areas' led to the concrete housing problems. Together with the technological problems, the mass movement of peoples from stable communities introduced unfamiliarity and uncertainty. In many instances the standard of accommodation provided in the 1960s was better than the two storey brick nineteenth century housing. Unfortunately the levels of comfort were soon found to be not much better than before; weather penetration of walls, condensation, difficulty in heating rooms, unsatisfactory services. The combination of poor housing standards, social doubts and also changing perceptions has made concrete structures unattractive.

Another social concern connected with modern housing structures is exemplified by the debate over steel framed houses. These were built mainly in the 1940s and 1950s. A simple two storey frame, with steel framed roof (or in some cases timber) with upper floor clad in steel profiled sheeting, with a block and rendered ground floor, is a typical construction. These houses have been investigated by the Building Research Station and found to be structurally sound. Evidence of corrosion to some steel columns was found, but rarely. Therefore they are deemed to be no problem. There is evidence of corrosion occurring to the external cladding. Insulation to the walls is negligible and consequently they can suffer from condensation and are difficult and expensive to heat. Many of these houses have been bought from local authority owners by the sitting tenants. Now, in many estates over the UK, they are finding difficulty in selling their houses. Building societies are wary of lending money for their purchase, not necessarily due to distrust of the long term structural safety, but because of their deterioration. It has been calculated that refurbishment would be at least £30,000, in 1990 prices. When this is set against present and future value it can be 70% of the total. For example, when refurbished in certain areas of the UK the house would only sell for £40,000 to £45,000. Therefore by the time that the cost of the refurbishment has been recovered and any outstanding mortgage on the property from the initial purchase, very little money is left to contribute towards the purchase of another property. There are some estates where the houses are virtually unsaleable and are therefore blighted. The social consequences of this are far ranging.

The point being made is that when considering the options for solving the problems of defective and/or sub-standard buildings, the wider social issues do need to be considered by technologists. Solutions are not derived from a black hole containing only technological items. They are developed with the needs and aspirations, and priorities, of society, whether it is for public benefit or private gain.

WARNINGS

Refurbishment has not been taken very seriously in the past. It has been associated with repairs and maintenance and carried out, in the main, by small company builders and specialist consultants and designers. Wherever awards have been given, new architectural styles applauded, and innovative constructions adopted, they have largely been on new build. Research programmes in construction management are primarily slanted towards new build activities. Product research and development has been directed towards markets in new structures. The pace and glamour of designing has been in the direction of creating new buildings. An analysis of subjects and topics taught in built environment degrees emphasises new build. Very little construction technology or architectural design is taught in relation to maintaining, repairing or refurbishing buildings. Yet in the UK this sector is virtually half the gross expenditure on buildings.

This is reflected in the quality of design and construction applied to maintaining the built environment. There is much anecdotal and experiential evidence to show that defects are inadequately diagnosed, repairs are wrongly specified and solutions poorly executed. Some major problems can occur, such as the overcladding peeling off as shown in figure 9.6.

Concerns are being raised about the integrity of austenitic stainless steel. This is widely specified in fixings tying claddings to the structure. Stress corrosion has been found which leads to the sudden fracture of the metal under load.

Overcladding systems, whether directly applied laminates or rainscreen are used to give protection and thermal enhancement to inadequate claddings. Although many systems have been utilised, some in places for up to ten years, there are question marks regarding some performance claims. For example, there is little evidence to justify the claim that a pressure equalisation is just that. In that the cavity air

9.6 Failure of a laminate overcladding system. *High rise flats, Sandwell*

pressure behind the rainscreen is the same as that prevailing externally. The idea being that gaps can be left in the screen to allow the free flow of air behind into the cavity. This does not produce disequilibrium in the pressures which could cause entrapment of rainwater. Another phenomenon requiring investigation is the presence of interstitial condensation in the cavity and between the different layers of a system. Initial on-site research shows that levels of condensation are higher than expected.

On one directly applied laminate overcladding system horizontal cracks of observable magnitude are occurring. Why? Does this portend serious trouble? Can they be sealed effectively?

Problems of workmanship in carrying out basic repairs and reinstatement are still prevalent. Brick/block leaves in cavity walls used in a PRC reinstatement scheme were observed to be built in different gauges. This manifested in horizontal joints not being at the same height. The wall ties were placed so that the highest level was in the external brick bed joint – the ties sloped towards the inner block leaf.

Concrete repairs have been known to fail within a short time of the remedial work. Standards of workmanship are erratic. A conscientious main contractor has had to remove operatives from class A repairs even though they purported to have been trained.

A debate rages amongst specifiers as to the best method of holding ties into existing backgrounds. The systems are not in doubt but the integrity of the installation process causes concern. This mainly applies to resin bonding type of ties. A two pack system demands very careful mixing at time of installation. A one pack continous system demands that the correct amount of fluid is added. It was noted on one contract that the site manager had to pre-measure and place into sealed containers the exact amount. This was taken up by the operatives in the access cradles with strict instructions to use all the measure with the dry material. This had to be invoked after operatives were seen not to be using the measuring flasks correctly. Also, if some was spilt, it meant less went in. Who was going to bother returning to the ground from 15 storeys just to get a little drop of water?

What is needed is much more research and development into all aspects of the maintenance, repair and refurbishment of buildings. Much good work is already going on, mainly by individual product manufacturers. They have scant control over where the product is used, how it is used in combination with other products, and the conditions in which it is installed. Specifiers and supervisors are awakening to the need to be more circumspect in design and installation, but much more needs to be learnt about existing systems and carried forward into developing repair and refurbishment systems which will not themselves fail prematurely.

It is because relatively young buildings, due to design and construction faults, are failing that much repair and refurbishment work is being carried out. Of course any building requires periodic and effective maintenance, but many of those buildings constructed with concrete, steel, timber and incorporating a variety of claddings are prematurely failing. Work is having to be carried out unnecessarily at a great cost. In many instances the building's users see no gain from

9.7 Careful detailing and workmanship are required to ensure a watertight joint between a reinstated house and a PRC house. *Wates houses, Wolverhampton*

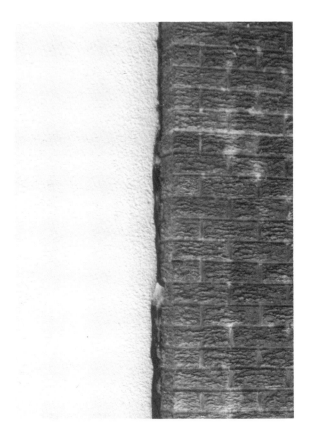

IS IT WORTH IT?

the expenditure. It is therefore of great importance that the remedial and improvement works are correctly designed and carefully executed. One last salutary story relating to the failure of modern buildings. A hospital building with in situ concrete frame, with glass reinforced concrete double sandwich cladding panels has failed after only six years, being constructed in the early 1980s when technologists should have been aware of possible problems. The claddings are precast GRC with a combination of outer GRC, insulation, middle GRC, another layer of insulation with an internal skin of GRC. The outer GRC skin takes all the environmental impact. It expands and contracts, freezes and thaws, is pulled and pushed. Its thickness and properties are such that it is constantly moving. This movement is excessive and is causing delamination, cracking and ingress of water. The building is to be overclad using a lightweight aluminium rainscreen system at a current cost of £1.7 million. It is to be hoped that this will perform adequately and with longevity. But it is £1.7 million which should have been spent by the building owner. There are many lessons still to be learnt.

SUMMARY

A reminder of the course this book has taken. A full circle has been drawn as we started with problems and end with a warning about further problems. After the initial introductory chapter the problems were described. Not only were the technical problems

9.8 Cladding. *Grande Synthe, Dunkerque, France*

relating to material deterioration, unsuitable combinations and poor workmanship discussed, but reference was made to, in some building types, a complementary social problem.

The need for a systematic and stage-by-stage process of investigation and diagnosis was stressed. A combination of checking initial design calculations and drawings, opening up, probing and testing materials is required to produce a picture of how the building is performing.

When deciding what action to take with respect to maintenance and remedial work, the future of the building takes prominence. On the technical front a fairly clear cut opinion could be determined with respect to long, short or medium term use. Techniques such as life cycling costing and value engineering can help. The social and economic factors have to be brought into the option process. As people, especially with respect to housing, have a greater awareness of the issues involved, they will take a more contributary role in deciding what is best for their homes.

In order for the decision to be carried, clear and comprehensive documents need to be produced. A balance will have to be struck between prescriptive specification and allowing design/build solutions. In the latter case warranties and meaningful guarantees will be necessary from the suppliers. The greater introduction of quality management systems, together with standards on workmanship, will go some way in improving the confidence levels in obtaining conformance to specification.

The bottom line of any project is cost. Costs can be directly determined by the state of the building and they can also be influenced by the specification and method of repairs. The contractor can also influence costs by considering the process of the works. What must also be included in the costs is the future maintenance and running expenses.

In some repair schemes the process and sequence of work is laid down by the specification. On site the contractor is responsible for carrying out the works in accordance with the specification. The experience of good contractors can make the repairs more efficient and effective. The greatest problems arise when the contractor has to work on occupied buildings. It is common to find that well over 50% of management time is spent on liaising with the occupiers.

Chapter 8 presented a number of studies showing how remedial and refurbishment work has been carried out to a variety of buildings. There is no one right answer. Each study has to be seen in its own set of circumstances but there are some general principles. These principles were described in the earlier chapters and should be seen against the given examples.

This last chapter opens with the question of whether refurbishment is worth it. On many measurable counts it is. Note should be taken that failures have occurred to remedial works and warnings have been given.

This book can only be an introduction to the activity of refurbishing modern buildings. The principles and issues have been described and discussed. There is no doubt that much more needs to be researched, developed and disseminated. If this book generates further interest and consequent action in ensuring effective refurbishment then it will have made a modest but useful contribution.

References

Association of Metropolitan Authorities, Defects in Housing Parts 1 and 2 1985

Baba, A, 'A proposal for a general system for making diagnosis of reinforced concrete buildings', Building maintenance & modernisation worldwide, ed L K Quah, pp 708–717, CIB/National University of Singapore, Longman 1990.

Baba, A, 'Overview of repair and renewal methods for reinforced concrete buildings in Japan', Building maintenance & modernisation worldwide ed L K Quah pp 852–860 CIB/National University of Singapore, Longman 1990

Baba, A, Senbu O and Matsushima, Y, 'Effect of various surface layers on the life span of reinforced concrete buildings', Building maintenance & modernisation worldwide, ed L K Quah, pp 708–717, CIB/National University of Singapore, Longman 1990

Baldwin, G R, 'The impact of climate on the durability of reinforced concrete', Building maintenance & modernisation worldwide, ed L K Quah, pp 688–697, CIB/National University of Singapore, Longman 1990

Bonshor, L L and Harrison, H W, 'Rehabilitation – a review of quality in traditional housing', BRE Report 1990 BR166

Bonshor, R B and Harrison, H W, 'Quality in traditional housing Vol 1: An investigation into faults and their avoidance' Building Research Establishment 1982

Boon, J and Robertson, G, 'Coping with mid-life crisis in buildings – a market driven decision model for addressing the problems of building modernisation, refurbishment and rehabilitation' Building maintenance & modernisation worldwide, ed L K Quah, pp 483–492, CIB/National University of Singapore, Longman 1990.

Brandon, P S, 'Life cycle appraisal – further considerations', *Building Maintenance Economics and Management*, ed Alan Spedding, Spon London 1987

Briggs, T, Renewal theory and its practical spreadsheet application for the determination of optimum replacement strategies Building maintenance & modernisation worldwide ed L K Quah, pp 224–233, CIB/National University of Singapore Longman 1990

Chandler, I E, *Building Technology* Vol 2: *Performance*, Mitchell, London 1989

Chandler, I E, 'Economic appraisal of reinstatement repair methods to PRC type houses', CIB W55 Building Economics Conference Sydney Australia 1990

Coleman, A, *Utopia on Trial. Vision and reality in planned housing*, Hilary Shipman, London 1985

Coskunoglo, O, and Moore, A, 'A decision model for the building renewal problem' Building maintenance & modernisation worldwide, ed L K Quah, pp 214–223, CIB/National University of Singapore Longman 1990

Currie, R J, Armer, G S T, and Moore, J E A, 'The structural adequacy and durability of large panel system dwellings Part 2: Guidance on appraisal' Building Research Establishment 1987

Davenall, A, 'Feasibility of refurbishment of Ronan Point', Newham London Borough Council (1985)

de Vekey, R C, Ballard, G, and Adderson, B W, The effectiveness of radar for the investigation of complex LPS joints', Proceedings of the Life of Structures (the role of physical testing) conference Brighton UK April 1989

Dell'Isola, A J, *Value Engineering in the Construction Industry* 3rd edition, Van Nostrand Reinhold 1982

Flanagan, R and Norman, G, 'Life cycle costing for construction', Royal Institution of Chartered Surveyors, London 1983, reprinted 1989

Gilleard, J, 'The value of Value Engineering' *Building Technology & Management*, October/November 1988

Gow, H A, 'Sinking funds and housing asset management' Building maintenance & modernisation worldwide ed L K Quah, pp 259–265, CIB/National University of Singapore, Longman 1990

Harper, D, *Building: The Process and the Product*, Construction Press, London 1978

Kelly, J and Male, S, 'Value management and quantity surveying practice', *Chartered Quantity Surveyor* October

1987, pp 37 and 38

LAU, L C, Refurbishment: an evaluation of financing practices in Singapore, Building maintenance & modernisation worldwide, ed L K Quah, pp 266–277, CIB/National University of Singapore Longman 1990

McCABE, S, 'Low rise steel and timber', TERN Project, Birmingham Polytechnic 1988

MOTTERSHAW, T, 'Evaluation techniques', TERN Project, Birmingham Polytechnic 1988

NEVILLE, A M, Creep of plain and structural concrete, London Construction Press 1983

NEVILLE, A M, Properties of concrete, 3rd edition, Longmans Scientific & Technical 1986

NEWMAN, O, Defensible space, Macmillan, New York 1972

NIERSTRASZ, F H J, 'Some economic aspects of maintenance, modernisation and replacement', Proceedings of Rotterdam Seminar CIB W70 Research on Maintenance and Modernisation, CIB publication 54 1980

PSUNDER, M and ZAJA, M, 'Maintenance, management of concrete buildings', Building maintenance & modernisation worldwide ed L K Quah, pp 809–812, CIB/National University of Singapore, Longman 1990

RIDLEY, P 'Evaluation and repair', TERN Project, Birmingham Polytechnic 1988

RUSSELL, B, Building systems, industrialisation and architecture, Wiley 1981

SPRING, MARTIN 'Projecting Euston's new image', Building 1 September 1989 pps 34–39

STROUD, C, 'Dismantling of Ronan Point: The Feedback', Newham London Borough Council (1984)

TUCKER, S N and RAHILLY, M, 'Life cycle costing of Housing Assets' Building maintenance & modernisation worldwide, ed L K Quah, pp 162–171, CIB/National University of Singapore, Longman 1990

WARN, BURGER, 'A working model for the choice of modernisation measures', Working Commission W70, CIB Proceedings of Rotterdam Seminar Publication 54 1980

ZHAO, G F and LI, Y G, 'A criterion for evaluating the performance of existing building structures' Building maintenance & modernisation worldwide ed L K Quah, pp 679–687, CIB/National University of Singapore, Longman 1990.

INDEX

Numbers set in *italic* refer to illustration pages

Abortive fixings 88
Abseiling 48, 64, 78, 87, 88, 90, 107
Access 50, 83, 85, 87, 105ff, 114, 127
Access to equipment 47, 48, 50
  multistorey buildings 105ff
AFC Fram 63
Agrément Certificate 93
Aluminium mullions 159
  sheets 70
Analysis, material 79
  option 81
Appraisal 55
Architectural stylistic movements 9
Asbestos cement sheets 26
Ashford 125
Association of Metropolitan Authorities 12
Assurance testing 99
Atholl system 42

Baffles 25
Balcony panel, removal of *106*
  slabs 95
Bar chart 102
Bathroom extension *152*
Bemis, Albert E, *Rational Design* 9
Bills of approximate quantities 102
Bills of quantities 78, 83, 112, 138
Birmingham 55ff, *82*, *85*, *107*
  Bull Ring Shopping area 64
  Castle Vale *62*
  City of, Housing Department 52, 63
  college building *127*
  Erdington *29*
  high rise flats *81*, *82*, *84*, 87, *115*
  high rise housing 60, *101*, *109*, *110*
  multistorey flats *114*
  offices 13, *107*
Birmingham Polytechnic, TERN Project 35, 42
BISF 11
Bison cross wall, Walsall 27, *135*, *136*, 137
Bison houses, 11, *26*, *126*
Bison wall frame system 38, *41*
  checklist 39
Boot houses 10
Boroscope 47
Boston Redevelopment Authority 12
Boswell type house 64, 160
Brick cladding 13, *15*
  replacement *20*, 98, *127*
Bristol, reinstated house 10
Brisol, Wates houses *113*
British Standard Specification 93
British Standards Codes of Practice 57
British Steel Construction (BISF) 11
Building Research Establishment (BRE) 33, 42, 55, 69
  reports 11, 48, 51
  station 165
Buildings' users 30, 55, 80
Bulging 26

Carbonation 22, 40, 41, 91, 93
  chemistry of 22
  drilling and testing for *23*, 51
  of concrete 15, 21, 95
Case studies 134 *et seq*
Central heating systems 20, 151
Certificate of Handover 85
Chandler 55, 70, 147, 163
Chemical attack 40
CIBSE 164
Chicago: housing 15, *31*, *118*, 154–57
  apartment building *155*
  external elevator tower *71*
  offices *162*, 163
  satellite access 97
Chicago Housing Authority 13
Chloride attack 21, 23, 24
City of Birmingham Housing Department 52, 63
Cladding components 10
  leaf 41

system testing 70
Claddings 9, 13, 26, 62, *168*
  brick 12, 63, *155*
  steel sheet 29
CLASP 11
Clerk of Works 89
  reports 36, 56
Climate cycle UK 22
Code of Practice UK 23
Colemark 30, 32, 33, 65, 75
Collapse of building 57
'Colourend' finish 137
Comfort 60
Competition 89
Complaints book *122*
Computer analysis printout *53*
Computer based condition survey record *68*
Computers 52, 53
Concrete 11, 28, 33
  chemical defects in 17
  floor beams 43
  repairing and finishing 91, 92, 119
  toe *115*
Concrete block Boot houses 10
Concrete blocks 10
  cracks 22
  creep 21, 26
  repair 92
  shrinkage 20
  spalling *22*
Condensation 33, 42, 46, 70, 72ff, 157
Consortium of Local Authorities Special Project (CLASP) 11
Consultants liaison with 114
Corrosion cracking 41
Corrosion of reinforcement 21, 23, 24
Corrosion to steel framed structure 42, 46
Cost, capital 18, 160
Cost evaluation 161
Cost factors 105 *et seq* 169
Costing 109
  examples 116, 117

173

INDEX

Cracking in concrete 22, 26, 57, 73
Cradles 83, 85, 86, 87, 105, 106, 108, 109, 167
Creep 21
Crest Homes 145
Crime against people 32
Curing 110
Curtain walls of glass 28
Czechoslovakia, overcladding 14
    Prague housing 14, 15, *162*, *163*, *165*

Damage to buildings 32, 65
    vandal 31
Damp proof course 27, 199, 138
Damp-proof membrane 38
Davenall 13
Defects in design 34
Defects in structure 18, 78
    nature and extent of 34
    to brick panels *25*
Defects Liability Period *87*
Dell, 'Isola' 69
Delta bronze 24, 41
Desk study 36, 55, 56
De Vekey 69
Dilapidation schedule 120–122
Dimensional problems *133*
Distortion 28
Document, content of 83, 90, 92
Documentation 78 *et seq*
Documents, inspection of 123
Dorlonco, panels 10
Dust-proof screens 98

Economic evaluation 18
Economic life of building 18
Economic problems 14
Elemental repairs 88
    bills 90
Emergency 61
Endoscope 46
Energy 50, 74, 82
    evaluation 164
Epsicon system 145
Environment, control 33, 74
    improvements 76
Erection procedures, sequence of 124, 125
Evaluative techniques 73
Excrement 31
Extensions 16
External brick panel, replacement 99
    finishings 99

Factory produced units 12
Factory systems 9
Falling debris 57
Federal Housing Administration 12
Finishes 'colourend' 137
    flaking 21
    mosaic 28, 40, 41, *82*, 132, 139
    spalling 21
    surface failure 28
    wall 50, 80

Fire resistance 34
Fittings 50, 147, 151
Fixings 40, 46, 88, 125, 137
    abortive 88
    corrosion of 24
    into concrete 33
    panel 40
    poor 24
Fixture 50, 147, 151
Flanagan 66
Floor coverings 50
Floors 43, 69, 153, 154
Flow chart, programme for determining remedial work 35
Forms of tenure 15
Form of Contract 87, 89, 92, 99
Framed steel clad houses 42
France: housing 16, 17, 134, 149–54
    Chateaubriand, Savigny-sur-Orge *152*, 163
    Grande Synthe, Dunkerque *16*, *168*
    Gravelines *132*, *153*, *154*
Freeze/thaw cycle 28
Functional life of building 18, 160

Gallows brackets 159
Gas 50
Gillierd 69
Glasgow housing 13
    high rise, fenced off area *80*
    Pollockshaws, rainscreen panels *71*, *72*
'Global warming' 164
Graffiti 31
GRC 168
Grit blasting 92, 95, 105, 106, 132
Gullies 131
Gustafsson 66

Hammer test 93, 119
Harper 55
High rise buildings 9, 11, 13
Hospital building 168
    Leicester *20*
Housing and Urban Development Department (HUD) 12
HUD 12
Hull, UDG maisonettes *143*, *144*, *145*, *146*

Infill panels 10
In situ concrete
    floor 12, 43
    frames 9, 11, 12, 13
    precast 10
    ring beam 42, 43
    walls 12
In situ corner columns 44
Inspection
    cavity 46
    check list 37, 38, 39, 47
    external 44
    internal 45
    report 78
    services 50

    site 56
Inspection record, computer based 52
Inspecting, recording 51
    site 56
Institute of Civil Engineers (ICE) 102
Institute of Structural Engineers (ISE) 57
Insulation 28, 30, 46, 168
    cavity *20*
    low standards of 28
    'mineral wool' 46
    polystyrene 38
    sandwich 159
    styrofoam 145
    to walls 92
Intelligent buildings 74
Investigations 36

Joint Contract Tribunal (JCT) 102
Joints between panels 40, 41
Joints, design 26, 27

Kelley 82
Kitchen units 50, 129, 145, 155

Laboratory analysis 58
Large panel systems (LPS) 9, 11, 13, *14*, 50
    window panel *16*
Le Corbusier 9
Legionnaires disease 28
Lift maintenance engineers 20
Life cycle analysis 66
    cost management 66
    costing 66, 77, 169
    costs 81
    planning 66
Lift tower during construction *131*
Lifts 28, 155, 159
Lintel, timber 45
Litter 31
Loadings, speed of 24
Local authorities 56
Loft insulation 46
Loft space 46
Log Book 36, 51, 55, 78, 103, 104
London 13, 164
    Doddington Estate 18, *19*, *131*, *161*
    Edgware Road 157, *158*, *159*
    Office buildings, defective curtain walling 74
    multistorey office *164*
Low rise buildings 11, 108
    Reema houses 11
Low rise PRC houses 97
    timber frames houses 9
    structures 36
LPS 9, 11, 13, 16, 36, 50, 61

Maintenance cost 60
    history 36
Mastic 26, 27, 42, 55
    sealant 70

INDEX

Materials 109, 110, 111
   comparison of 108
   life of 69
Mesh grilles *31*
Metal frames 27
McCabe 42
Methodologies, comparison of 105
Mirrors 47
Modern buildings, definition 9
Mosaic finishes 28, 38, 40, 41, 62, 91, 132
   falling 61
   patched *82*
Mottershaw 34ff
Multi Disciplinary Team (MDT) 63
   Risk code 36

Nails, plasterboard 30
National House Buildings Council (NHBC) 98
Newman 31
Niernstrasz's formula 161, 163
Non-continuance of work 86
Non-traditional dwellings 10, 11
Nuneaton
   erratic tiles *45*
   rainwater penetration *30*
   removal of brick skin *95*

Obsolescence 18
   economic 74
   factors 81
   fixtures and fittings 151
   functional 74
   legal 75
   physical 74
   technological 75
   social 75
Occupied buildings 119
'Operation Breakthrough' 12
Option criteria 65
Options 60 *et seq*
Orbit houses 64
Order to Recommence 85, 86
Outdated facilities 13
Overcladding 100, *103*, 107, *149, 153, 157*, 166, 168
   failure of laminate *166*

Palliative reparts 63, 64, 83, 102, 107
Panel cladding 13
   falling off *13*
Panel systems 11, 13, *14*
Panels
   brick 11
   damaged *162*, 163
   distortion of 21, 28
   precast concrete 11, 13, *15*, 16, 24, 25
   sandwich 28
Parapets 24
Parkinson PRC House 67, *67*, 68, 69
Paris, La Courneuve *61*, *149*, *150*, 151
Party wall, removal of *147*

Payment for repairs 86
Performance criteria 73
   Specification 82
Phenolphthalein test 43
Plastic sheeting *15*
Points to investigate *40*
Policing, beat 77
Poor workmanship 13, 33
Portsmouth, common areas *32*
Power 75
PRC Cornish 10
PRC Homes Limited 98
Prime Cost of Daywork 86
Private developers 15
Problems 42
   peculiar to 'new methods' 13
   physical 18
Programmes 102, 113, 123
Protecting against falling mosaic *29*

Quality Assurance 27, 95, 102, 123, 125ff
Quality Control 20, 27, 57, 99, 110, 123
   inadequate site 27
Quality management systems 123, 125, 126

Radar 48
Radiography 48
Rainscreen cladding 72, 99
   overcladding 61, 71ff, 166
   System 71, *72*, 168
Rainwater 29
Rational Conversion (RC) Method 161
Record of work to be done 102, 104
Records 36, 51, 56
Records, limitation of 56
Reema houses 11, 42, *43*
   Hollow panel house 42
Refurbishment process 14
   Value of 17
Remedial solution failure 72
Remedial works, brief 82
Rent 15
Repairs, long term 92
   short term 90
Report, The 58, 59
Resident caretaking 76
Resin injection repairs 93
Ridley 12
Rising damp 29
Rochester, Cornish houses *98*
Ronan Point, London 13, 56, 57, 61
Roofs 69, 92
Rotunda, Birmingham 64
Russell 9

Safety 17, 34, 57, 79, 90, 100, 111, 127
   fire 50
Sandwell high rise flats *166*
Sanitary fittings 50, 151

Satellite platform hoists 107, *107*, 119
Scaffold 100, 105, 106, 108, 110, 119, 139
   access 32
   boards 100
Schindler Frame 9
Schindler, Rudolf (architect) 9
School buildings 29
Scotland
   high rise housing *80*, *89*
Pollockshaws, Glasgow *71*, *72*
   tenements in 11
   Tracoba system 11, 12
Sealants 21, 25, 26
Security grilles *31*, 155, *156*, 157
Services, building 18, 27, 159
   investigation 50
Sequence 108, 113, 128, 132, 140, 147
Sequencing 108
Sheet protection 100
Shopping precinct 18
Short term repairs 90
Single skin concrete walls 9
Site organisation 119 *et seq*
   methodologies, examples, 128 *et seq*
   visit 114
'Slum areas' 165
Social concerns 165
Social factors 78, 169
Societé Anonyne D'Habitations A Loyer Modre (HLMs) 16
Solihull housing *11*, *23*
Spalling 26, 28, 43, 57, 61
Spring 163
Stairs, repositioned *157*
Standard Method of Measurement 112
Stafford, high rise housing *106*
   in situ concrete frame *139*, *140*, *141*
   multistorey flats *111*
Steel 11, 23
   frames structures 9, 10, 29, 46
   rebars 23
   sheet cladding 29
   window frames 28
Steel poor reinforcement cover 13
Strategies, categories of 61
Structural engineer 83
Structural integrity 127
   report 80
Structures 20
   defects 21
Sulphage attack 98
Supervising Officer 83, 85, 86, 87, 88, 89
Surface finish failures 28
Sweden, life cycle cost equation 66
System building 10, 11
   techniques 11

Taunton, Airey houses *97*
   Woolaway houses *103*
Taunton Dene, PRC houses *91*

INDEX

Tender drawings 112
Tendering 109, 113
Tenants 56
Tenements in Scotland 11
Term Contracts 83, 102
TERN Project *35*, 42
Test cubes 112
Testing 48, 88, 111, 169
  site 56, 57
Tests common for modern
  buildings 48
  laboratory 58
Thermal insulation 42
  low level 13
  movement 26
Ties, corrosion of 21, 24, 41, 46
  'goal post' 41
  metal 48
  stainless steel floor *140*
  wall 24, *132*, 138, 140
Timber 11, 12
  framed structures 29
  houses 44
  inspection 44, 45, 46
  untreated 30
Tracoba system, Scotland 11, 12
Typical solutions 134 *et seq*

Ultra-sound 48
Uniformly distributed loads (UDLs) 24
UK: commerical 134, 157–59
UK: housing 134–48

Birmingham 13, *29*, 52, *60*, *62*, 63, *81*, *82*, *84*, 85, 87, *101*, *107*, *109*, *110*, *114*, *115*, *127*
Bristol 10, *113*
Hull 134, 142–46
Leicester *20*
Leicestershire *126*
London 18, *19*, 31, *74*, *76*, *131*, 157, *158*, *159*, *161*, 164, *164*
Midlands 160
Nuneaton *30*, *45*, *95*
Oxford 31
Portsmouth *32*, *125*
Stafford 134, 138–42
Taunton *91*, 97, 103
University of Sussex *22*
Walsall *79*, 134–38
Wolverhampton *167*
Unstable brickwork, replacement 98
Untreated timber 30
USA 17, 82, *see also* Chicago
  Southern California 9
User behaviour 33
  evaluation 55
  requirements 55, 75, 80

Valuation records 78, 103
Value engineering (VE) 69, 77, 169
  management team 82
Vandal damage 31
Vapour barrier 30, 42
Verbal instructions 86

Vertical platform hoists 105
Visual inspection 47, *53*
  testing 53

Wall ties 24, 27, *132*, 140, 167
  wire butterflies 24, 138
Walls, curtain 28
  external 10, *113*, 132
  finishes 50, 80
  insulation 92
  internal 44, 45
  loadbearing *10*, 45
  parapet *26*
Warn, Berger 161
Wates houses 128, *129*, *130*, 147
  Bristol *113*
  Leicestershire 126, 147, *147*, 148
  Nuneaton *128*
  Wolverhamptom *167*
Weather protection, ineffective 13
Welsh Office 75
Wimpey 'No fines' 11
Window frames, softwood 27
  steel 28
Windows 64, 70, 98, 149, 151
  double glazed 92, *135*, 151, 155
  reveals 44

YDG *143*, *144*, *146*

Zinc coating 24